Introduction to Advanced Manufacturing

Introduction to Advanced Manufacturing

SAE International

Warrendale, Pennsylvania, USA

400 Commonwealth Drive
Warrendale, PA 15096-0001 USA
E-mail: CustomerService@sae.org
Phone: 877-606-7323 (inside USA and Canada)
           724-776-4970 (outside USA)
FAX: 724-776-0790

Chief Product Officer
Frank Menchaca

Publisher
Sherry Dickinson Nigam

Director of Acquisitions
Monica Nogueira

**Director of Content Management/
Senior Production Manager**
Kelli Zilko

Production Associate
Erin Mendicino

Manufacturing Specialist
Christie Inman

Library of Congress Catalog Number 2019939938
http://dx.doi.org/10.4271/9780768093278

Information contained in this work has been obtained by SAE International from sources believed to be reliable. However, neither SAE International nor its authors guarantee the accuracy or completeness of any information published herein and neither SAE International nor its authors shall be responsible for any errors, omissions, or damages arising out of use of this information. This work is published with the understanding that SAE International and its authors are supplying information, but are not attempting to render engineering or other professional services. If such services are required, the assistance of an appropriate professional should be sought.

ISBN-Print 978-0-7680-9327-8
ISBN-MediaTech 978-0-7680-9096-3
ISBN-ePUB 978-0-7680-9329-2
ISBN-PRC 978-0-7680-9330-8
ISBN-HTML 978-0-7680-9328-5

To purchase bulk quantities, please contact: SAE Customer Service

E-mail: CustomerService@sae.org
Phone: 877-606-7323 (inside USA and Canada)
           724-776-4970 (outside USA)
Fax: 724-776-0790

Visit the SAE International Bookstore at books.sae.org

contents

CHAPTER 3

Subtractive Manufacturing   41

CHAPTER 4

Additive Manufacturing   57

CHAPTER 10

Smart Manufacturing   177

After having taught manufacturing at a university level for more than a decade in three different continents for a few thousand Mechanical, Industrial, Manufacturing, and Aerospace Engineering students, we noted a gap in available resources. On the one hand side, there are the comprehensive manufacturing handbooks, and on the other hand, there are high-level texts that are missing the depth and breadth, with respect to manufacturing basics, we feel every engineering student should be equipped with. This realization led us to develop a concise yet modern textbook that introduces students to state-of-the-art advanced manufacturing, covering the breadth of the field, while providing a sufficient depth without overburdening the students. From the inception of the book in the mind of the authors, the idea was constructed around what can be realistically covered in one semester in a seamless manner. We also wanted an out-of-the-norm, approachable book that talks directly to the student and that can be considered accessible – to get our readers excited about the fascinating field of manufacturing. Manufacturing textbooks often fall into the trap of becoming "History of Manufacturing" where the covered material is partially outdated. We often joke in our classrooms that if the students fail, they might learn different materials the next semester! While said in a joking manner, that summarizes neatly how much manufacturing changes and develops constantly, moving forward to unlock innovation to service communities.

This book represents a point in time, like all engineering projects, where we decided to release it! There are tons of ideas that we would have liked to include as well, but the goal is to remain an approachable textbook without falling into the trap of a 1000-page textbook that becomes unaccessible both material-wise and financially to the students.

The book is separated into 10 chapters: (1) Introduction to Manufacturing, (2) Deformative Manufacturing, (3) Subtractive Manufacturing, (4) Additive Manufacturing, (5) Assembly Processes, (6) Computer Aided Manufacturing (CAM), (7) Polymers Manufacturing, (8) Composites Manufacturing, (9) Manufacturing Quality Control and Productivity, and (10) Smart Manufacturing.

Chapter 1, **Introduction to Manufacturing**, presents a general introduction to manufacturing families, manufacturing evolution, and materials in manufacturing. *Chapter 2*, **Deformative Manufacturing**, introduces deformative manufacturing including both forming and casting processes. The chapter further details rolling, forging, extrusion, and sheet metal processes. The chapter also details sand casting and die casting. *Chapter 3*, **Subtractive Manufacturing**, develops subtractive manufacturing topics such as milling, turning, and drilling. *Chapter 4*, **Additive Manufacturing**, introduces the reader to additive manufacturing key technologies such as fused filament fabrication, selective laser sintering, and stereolithography. *Chapter 5*, **Assembly Processes**, details assembly processes such as mechanical joining and welding. We introduce CAM in *Chapter 6*, **Computer Aided Manufacturing**. CAM is introduced in the context of the numerical chain with a detailed understanding of the mathematical foundation of CAM, manufacturing references, and generation of G-Code. *Chapter 7*, **Polymers Manufacturing**, specifically introduces deformative manufacturing of polymers manufacturing such as extrusion, injection, and blow molding. This chapter also covers fundamentals of polymer materials in the context of manufacturing. *Chapter 8*, **Composites Manufacturing**, is tailored toward composites manufacturing, a subset of additive manufacturing. We introduce automated fiber placement, a prominent and rapidly developing composite manufacturing process, and illustrate the basics of composite materials in the context of manufacturing. *Chapter 9*, **Manufacturing Quality Control and Productivity**, focuses on manufacturing productivity and quality, with topics such as cycle time, cost estimation, design for manufacturing, inspection methods, statistical tools, and design of experiments being introduced. The book ends with *Chapter 10*, **Smart Manufacturing**, on innovation in the context of Smart Manufacturing.

This book would have not been possible, if we did not get the manufacturing education we got, through these cornerstone and foundational books:

- Fundamentals of Modern Manufacturing: Materials, Processes and Systems; 3rd Edition; Mikell Groover; Wiley ISBN 0-471-74485-9

- Introduction to Manufacturing Processes; 3rd Edition; John A. Schey; McGraw Hill; ISBN 0-07-0314136-6

- Manufacturing Processes and Systems; 9th ed.; Philip Ostwald & Jairo Munoz; Wiley ISBN 0-471-04741-4

- DeGarmo's Materials and Processes in Manufacturing; 10th Edition; JT Black & RA Kohser; Wiley ISBN 978-0-470-05512-0

- Manufacturing Processes for Engineering Materials; 5th Edition; S Kalpakjian and S Schmid; Pearson ISBN 978-981-06-7953-8

The above books, although from the previous decade, represent the books we nurtured our manufacturing knowledge through as students and early in our careers when we were assigned to teach the fascinating topic of manufacturing.

Moreover, we would like to highlight the companion website introtomanufacturing.com where multiple resources are made available to students. Additional exercises, YouTube video links, online quizzes, and other resources are offered to facilitate the overall student experience. We encourage faculty to reach out to us at ramy.harik@gmail.com for additional materials, such as exam samples. We have a faculty mailing list that provides detailed support to any faculty that adopts our textbook for their classrooms!

Also, we would like to propose the following outline for faculty adoption in teaching manufacturing. We propose to split the class into four cycles:

- Cycle 1: Introduction and Deformative Manufacturing (Chapters 1 and 2)

1.	Manufacturing Families	1
2.	Manufacturing Evolution	1
3.	Manufacturing Materials	1
4.	Deformative: Forging	2
5.	Deformative: Extrusion	2
6.	Deformative: Rolling	2
7.	Deformative: Sand Casting	2
8.	Deformative: Die Casting	2
9.	Deformative: Sheet Metal	2
10.	Exam	1
11.	Solution Exam	1

- Cycle 2: Subtractive and Additive Manufacturing (Chapters 3 and 4)

12.	Subtractive: Introduction	3
13.	Subtractive: Fundamentals	3
14.	Subtractive: Turning	3
15.	Subtractive: Milling	3
16.	Subtractive: Drilling	3
17.	Additive: Introduction	4
18.	Additive: Fused deposition modeling	4
19.	Additive: Selective laser sintering	4
20.	Additive: Stereolithography	4
21.	Guest Speaker	
22.	Exam	2
23.	Solution Exam	2

We propose that Manufacturing classes have one expert guest lecture, as well as one field trip to a real manufacturing industry.

We wholeheartedly thank those who supported this book through reviews and critical feedback. First and foremost, our students at the University of South Carolina, Lebanese American University, and West Virginia University who provided detailed and honest feedback on the materials and delivery over the last semesters while we developed this book and accompanying lecture materials – without them, we could not have achieved the quality of this first edition. Furthermore, we thank our esteemed colleagues and friends Clint Saidy, Dawn Jegley, Lucas Hof, Marc Haddad, Max Kirkpatrick, Royal D'cunha, and Todd Hamrick, who's critical feedback, reviews, and expert insights have been invaluable during the creation of this book. Support was also received from Georges Fadel and Juergen Lenz.

From both the authors, we would like to dedicate this work to our families: Leona, Celia, Jonas, and Samar Harik; and Dominic, Sofia, and Irene Wuest. This took a substantial amount away from our immediate families, hoping that it would serve our extended scientific family and students!

Ramy Harik, Thorsten Wuest

1

Introduction to Manufacturing

Manufacturing and the manufacturing industry is considered the backbone of every economy, whether in developed or developing countries. Countries with innovation in manufacturing are better equipped to counter the effects of economic crises leading to a faster return to economic stability. A survey of developed countries demonstrates that they are viable based on (1) level of industrialization and/or (2) availability of natural resources. Economists often state the comparably low but certain return of investment in manufacturing. According to the National Association of Manufacturers, manufacturing has the highest multiplier effect of any economic sector: For every US dollar spent in manufacturing, another $1.89 is added to the economy (2017 data). Furthermore, the total multiplier effect of manufacturing is estimated to be up to $3.60 for every dollar of value-added output, which roughly translates into 3.4 jobs created by every manufacturing worker (NAM, 2018).

The term "Manufacturing" no longer reflects its historical name origin but has developed over the centuries alongside social and technical developments. With its Latin origin, manufacturing stems from the words *manus* (hand) and *factus* (make). Nowadays, the term manufacturing incorporates a multitude of definitions that, even though they have different meanings, refer to a (series of) process(es) enabling a design to become a physical tangible product. On the most abstract level, manufacturing transforms inputs into outputs while adding value to a physical product or part (see Figure 1.1). Each manufacturing process increases the value of a product or a part, which basically means, someone is willing to pay more for the product/part after the manufacturing process is concluded than before. If this was not the case, one would not perform the manufacturing process in the first place. This highlights the economic component that is always part of manufacturing. Recent trends refer to Manufacturing 4.0 (or Industry 4.0 in Europe and Smart/Advanced Manufacturing in the USA) to highlight the current era of manufacturing.

FIGURE 1.1 Transformation model in manufacturing and value creation (Wuest, 2015).

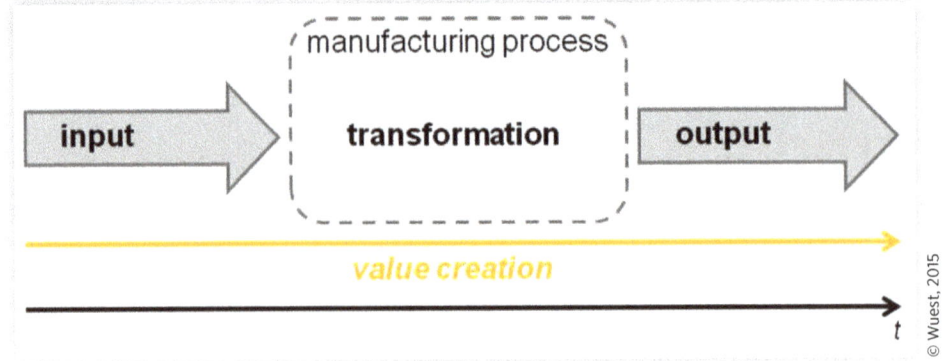

© Wuest, 2015

In this book, we aim to provide a comprehensive overview of the manufacturing domain itself including its main processes and technologies used. Furthermore, we will inform the reader of manufacturing's inherent connections with design, materials science, and other closely related domains. Our unique approach focuses on instilling a solid understanding that manufacturing is not an independent field but that manufacturing and design are intertwined as two sides of the same medallion. Manufacturing engineers and designers need to work closely together to achieve the best possible solution. We believe that there is a reason why manufacturing classes are regularly associated with Mechanical and Aerospace departments at some universities, while at others, the Industrial and Systems Engineering departments are their academic homes. In this book, we serve both, students who are more process and manufacturing technology oriented and students who are mainly interested in engineering and product design. The former will be exposed to, e.g., the effects of design decisions on manufacturing, while the latter will, e.g., better understand what limitations the available manufacturing resources enforce on their design decision. We are strong believers that manufacturing and design need to better communicate and work collaboratively to achieve the best results and that our book will prepare the reader to be champions of this view in the future.

This introductory chapter contains three main sections: *Manufacturing Families*, *Manufacturing Evolution*, and *Manufacturing Materials*. The *first section* will present an updated viewpoint of manufacturing families and their classification: Deformative, Additive, and Subtractive manufacturing. The *second section* will stage the manufacturing evolution from the manual processes of the past to the current Manufacturing 4.0 concept and what that entails technologically (i.e., evolution of the wheel depicted in Figure 1.2). This will enable the reader to grasp the natural progression of manufacturing concepts and prepares her/him for subsequent advanced studies in manufacturing. The *third and final section* will introduce engineering

FIGURE 1.2 Illustration showing the evolution of the wheel starting from a stone wheel and ending with a steel-belted radial tire and the effect of advances in materials in manufacturing.

© James Steidl/Shutterstock

materials from the manufacturing perspective: Design desired properties naturally counter manufacturing process simplicity and cost considerations.

1.1 Manufacturing Families

In this section, we will present a structure of the broad field of manufacturing that is both easy to understand yet flexible enough to fit the diverse field of manufacturing comprehensively. After reading this chapter you should be able to classify every current and future manufacturing process and add it to one of the manufacturing families introduced below.

1.1.1 Notion of Product and Part

In manufacturing, we typically obtain either parts or products. Historically, parts were associated with manufacturing shaping and processing techniques, resulting in one continuous form of material. Similarly, products were associated with manufacturing assembly techniques, resulting in one product. Figure 1.3 demonstrates the aggregation of those elements:

- Part 1 was shaped using Shaping Process X.
- Part 2 was shaped using Shaping Process Y.
- Sub-Product 1 was assembled using Assembly Process Z on Part 1 and Part 2.

The motion conversion system is a sub-product of car engines, which are a sub-product of cars. The motion conversion system consists mainly of three parts: Piston, Connecting Rod, and Crankshaft (see Figure 1.4).

FIGURE 1.3 Aggregation of elements to form a product.

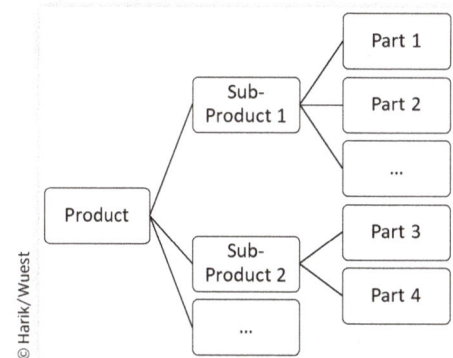

© Harik/Wuest

FIGURE 1.4 Assembled motion conversion system.

© Maxx-Studio/Shutterstock

1.1.2 Initial, Intermediate, Near-Net Shape, and Net Shape

We can understand manufacturing as a transformation process that adds value to a part or product. The value-adding portion is essential, as it justifies the investment of resources to transform the product. Without the manufacturing process adding value for at least one, the manufacturing process would not be conducted in the first place. You will notice that many times the term "Manufacturing" is used interchangeable with the term "Production." However, there is a distinct difference in the meaning of both when we take a closer look. On the one hand, manufacturing is connected to industrial production in most cases. Whereas it is true that every type of manufacturing is also production, not all production is necessarily manufacturing as production describes converting input to output in a broader, more inclusive sense. An example for a production process which cannot be described as a manufacturing process is a book. While the physical book (e.g., paper, printing, book assembly, etc.) itself can surely be manufactured, the content, the creative work (what is written in the book) cannot. Another common example is that services can be produced but not manufactured (see Figure 1.5).

Manufacturing of the majority of reasonably complex parts and products involves a succession of a variating number of different processes. Manufacturing processes in their multiple classifications enable a progression of materials from an initial shape to a finished shape as intended by the design. The different processes transform the initial shape into multiple intermediate shapes. Only when the shape is "almost" as intended by the designer, we classify the output as a near-net shape process. Once we perform the last (final) process, we classify the output as the net shape or the finished part. Several manufacturing processes are considered economically advantageous since they offer near-net shape or net shape output. Definitely, numerous criterions influence the selection of the manufacturing process, such as intended design properties and availability of manufacturing resources, therefore forcing the adoption of processes that results in intermediate shapes.

The same products and parts can be manufactured using a combination of different processes and machine tools. The manufacturing engineers and process planners have a certain degree of freedom in their decision, depending on the design and other restrictions. In most cases, the economic perspective comes into play when deciding the best way to manufacture a part or product. While for small batch sizes, such as specialty, highly complex products, an expensive manufacturing process such as Selective Laser Sintering makes sense economically and technically; for the same product in a larger production volume, a succession of casting, milling/grinding, and final assembly is often preferable from an economic standpoint.

Figure 1.6 illustrates the concept on the manufacturing of engine blocks made out of aluminum. Typically, we introduce raw materials, in the form of ingots, to a melting furnace to obtain molten aluminum. Then, we pour the molten metal into a casting mold to obtain a rough engine block. Following, we machine the contact surfaces, such as the cylinder walls, to obtain the finished engine block representing the design net shape. In certain cycles, we have multiple processes generating different intermediate shapes prior to obtaining the net shape finished part.

FIGURE 1.5 Manufacturing as a transformation process (Wuest, 2015).

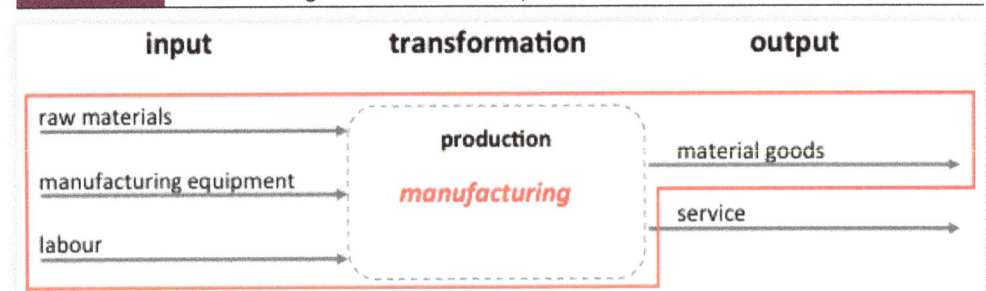

© Wuest, 2015

© Shutterstock/Harik/Wuest

FIGURE 1.6 The different shape classifications.

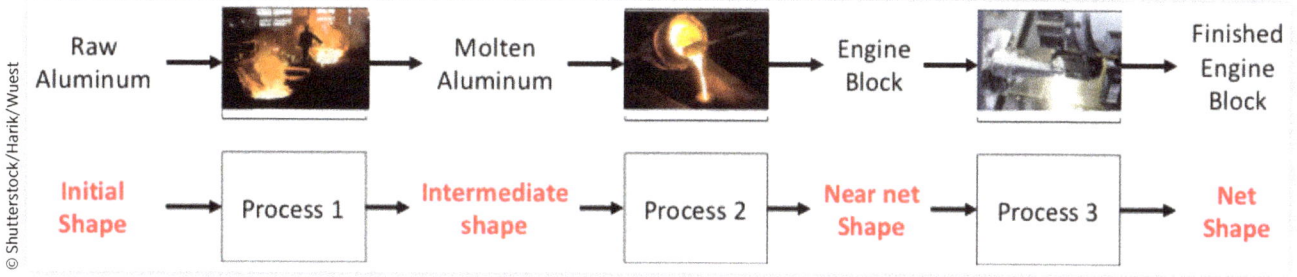

1.1.3 Classification

Numerous studies classify manufacturing processes into assembly - to obtain products - and processing, to obtain parts. This classification is rather obsolete since the status of direct digital manufacturing, most commonly known as Additive Manufacturing or 3D Printing, enables us to obtain a functional assembly with a single print out (= process). We offer a simple yet inclusive and scalable classification of manufacturing processes that is future proof into three main categories: *Deformative, Subtractive, and Additive.* We consider Assembly as a subsequent step. Figure 1.7 shows a selection of the principal manufacturing processes of each of the three defined categories. Each of the defined categories will be covered and detailed in Chapters 2, 3, and 4, respectively. Additionally, we cover two subsets of these categories in more detail: Polymers Manufacturing and Composites Manufacturing. Polymers Manufacturing is understood as a subset of Deformative Manufacturing, while Composites Manufacturing is considered a subset of Additive Manufacturing. We chose to address these two subsets in more detail because of their significance for industry, as well as for the rapid developments over the last years in these fields. Therefore, the reader will be provided with state-of-the-art insights in these novel fields with increasing relevance and impact on the manufacturing industry at

FIGURE 1.7 Classification of manufacturing families.

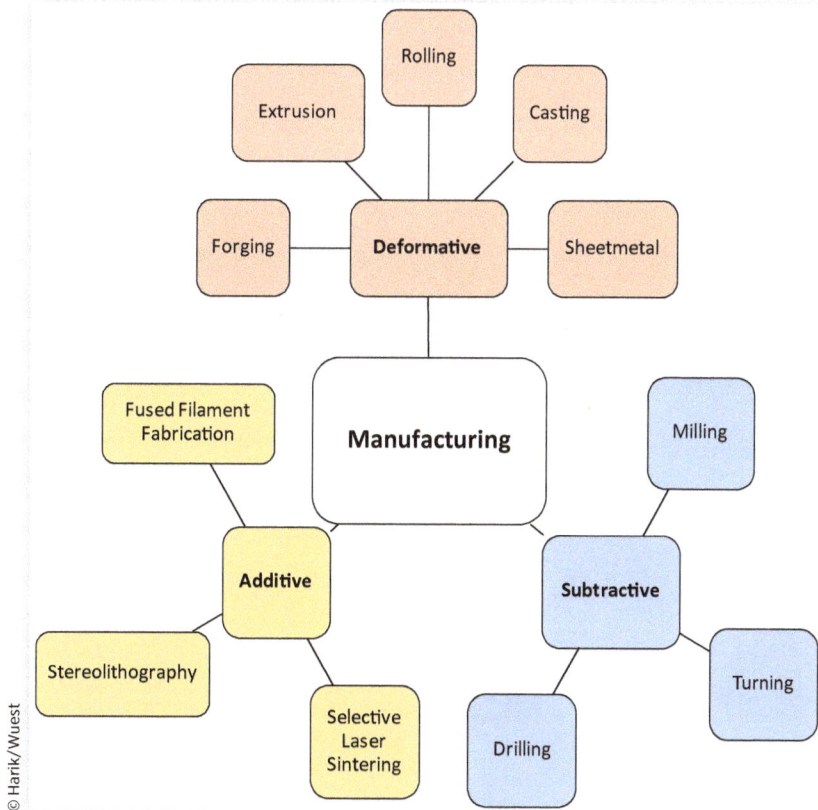

© Harik/Wuest

large. For certain industries such as aerospace, knowledge about these processes is even more important.

1.1.4 Deformative Manufacturing

Deformative Manufacturing represents processes where we transform the material from Shape A to Shape B without the addition or subtraction of material. The fundamental concept is that the volume of materials remains unchanged throughout the process. The transformation process can happen at different operating temperatures ranging from (below) room temperatures to ones above the material's melting temperature. We distinguish three major Deformative Manufacturing types: Forming, Casting, and Sheet Metal Manufacturing. Forming occurs at temperature operating ranges between room temperature and approximately 70% of the melting temperature value. Casting occurs at temperature operating ranges above the melting temperature. Forming requires the application of forces above the yield strength to generate a permanent plastic deformation. Forces can be applied to casting operations, particularly in Die Casting and in Polymer Manufacturing processes such as injection molding. We will present Deformative Manufacturing to the reader in more detail in Chapter 2 "Deformative Manufacturing" and Chapter 7 "Polymers Manufacturing." Chapter 2 will detail Deformative Manufacturing main processes: Forging (see Figure 1.8), Extrusion, Rolling, Sand Casting, Die Casting, and Sheet Metal Manufacturing. The latter is a combination of Deformative and Subtractive Manufacturing as it includes material removal processes such as blanking, punching, and shearing. Chapter 7 will illustrate Polymer Manufacturing, its materials and main manufacturing processes: Extrusion, Injection Molding, and Blow Molding.

1.1.5 Subtractive Manufacturing

Subtractive Manufacturing represents processes where we transform the material from Shape A to Shape B by subtraction of material. The fundamental concept is that we reduce the volume of materials throughout the process. Subtractive Manufacturing often generates superior structure quality of the product and/or part, compared to common processes from the Deformative and Additive Manufacturing families. Recent advances in Additive Manufacturing however suggest promising results that might change that historically accepted notion in the

FIGURE 1.8 The blacksmith trade of forging is a Deformative Manufacturing operation.

© Nejron Photo/Shutterstock

FIGURE 1.9   Jet fan milled on five-axis CNC machine.

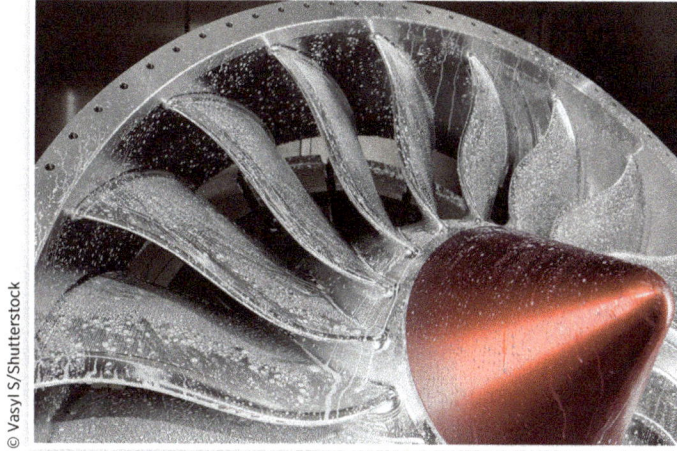

© Vasyl S/Shutterstock

future. Subtractive Manufacturing is not limited to traditional machining operations such as drilling, milling, and turning. All manufacturing processes that reduce the volume of materials are subtractive processes. Chemical milling (Industrial Etching) is one of those processes where we immerse a structure in a chemical bath, thus removing material of the part/product. We protect specific locations of the structure with shielding materials where no material removal is required. The final structure (and material removed) depends on the amount of time we immerse the starting structure in the chemical bath.

In Chapter 3 "Subtractive Manufacturing," this book will detail the traditional subtractive manufacturing processes, often referred to as *machining* in scientific literature, that are most commonly used in industry, including Milling, Drilling, and Turning. The result of a *Milling* process is a free-form or prismatic part. There are multiple configurations of milling machines available, such as five-axis, four-axis, and to this day the most common three-axis ones. These "axis" refer to the degrees of freedom/accessibility for the machine to perform the intended manufacturing operations. Generally, the higher the degree of freedom (= No. of axis) a machine is capable of utilizing, the more complex the parts/products can be manufactured with that machine. However, at the same time, the initial investment is significantly higher and the complexity of operating the machine requires highly skilled operators. The jet fan depicted in Figure 1.9 is an example of a complex product manufactured on a five-axis Computer Numerical Control (CNC) mill. *Drilling* is a particular milling type where the tool removes materials along one axis only. Therefore, we can use a five-axis Milling machine to drill a hole, while we cannot use a drilling machine to create a complex structure automatically. *Turning* results in a revolution part: a profile revolving around an axis. It is worthwhile to note the significant influence subtractive manufacturing exerted on Computer-Aided Design (CAD) and Computer-Aided Manufacturing (CAM). The Computer-Aided Manufacturing Chapter 6 "CAD/CAM" will provide more details on this later.

1.1.6 Additive Manufacturing

Additive Manufacturing represents processes where we transform the material from Form A to Form B by addition of material. The fundamental concept is that we augment the volume of materials throughout the process. Additive Manufacturing is currently under thorough investigation, and rapid progress is made on the technology, process, and materials. Methods to enhance the material properties and ways to increase the cost effectiveness are the principal research topics. Another topic that is heavily researched is the possible certification of parts/products manufactured additively. Additive Manufacturing offers numerous advantages as accessibility issues are no longer present. Challenges arise with respect to optimal building directions (in general) and fiber

FIGURE 1.10 Fused filament fabrication.

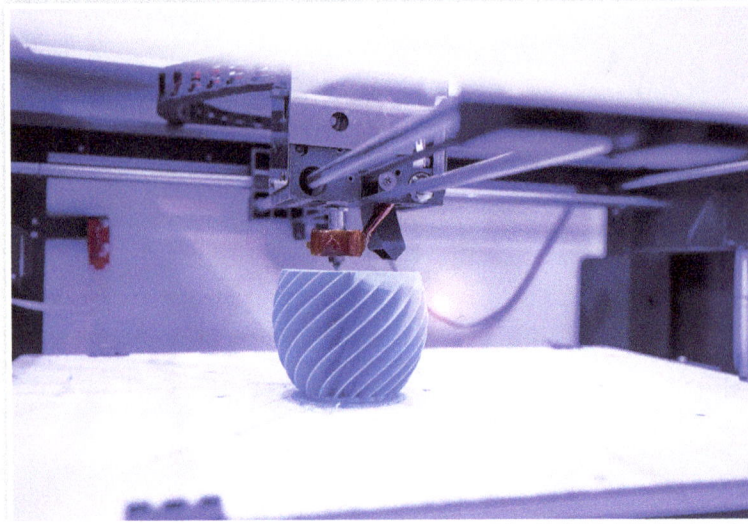

orientations (in particular with respect to composite manufacturing), as well as non-technical ones, e.g., in the field of Intellectual Property rights and liability. We separate Additive Manufacturing processes based on the physical form of the starting material. The most common starting forms are liquid monomers, filaments, and powders. In aerospace Composite Manufacturing, the common starting forms are unidirectional tape, unidirectional tows, and woven fabrics, typically impregnated with thermoset or thermoplastic resin system.

The book will present Chapters 4 "Additive Manufacturing" and 8 "Composites Manufacturing" that cover Additive Manufacturing comprehensively. Chapter 4 details the three main processes: Fused Filament Fabrication (see Figure 1.10), Selective Laser Sintering (SLS), and Stereolithography (SL). Fused Filament Fabrication uses a filament from a spool, transforms it into a molten state in an extruder head, and deposits it on a platform to generate the shape of the structure layer by layer. Different variations exist where either the extruder head is moving freely in space to deposit the material as directed by the digital instructions or the platform moves to ensure the correct material deposition, or a combination of both. SLS applies a laser power source on a bed of powder to sinter and create the desired structure. SL applies photopolymerization on a liquid basin of monomers to cure and generate the shape of the structure. All additive manufacturing processes use numerically controlled apparatus to control the deposition, sintering, and photopolymerization functions. Chapter 8 "Composites Manufacturing" highlights this novel field from two main perspectives: materials and processes. It will present a detailed overview of Automated Fiber Placement (AFP), the current state-of-the-art manufacturing process in manufacturing of highly sophisticated manufacturing goods, especially in the aerospace and aircraft manufacturing industry. The main structural components of Boeing 787 and Airbus 350 use AFP as their principal manufacturing process.

1.1.7 Assembly Operations

We obtain products (and sub-products) by performing assembly operations on separate parts. Assembly operations include mechanical joining, welding, and bonding. Mechanical joining requires shape modification of the original parts. Drilling is required to enable threaded fasteners or other components. This typically makes the assembly process costly. However, cost might ironically be the reason of having a product being designed as an assembly in the first place. There are several reasons for this: the main one is reduced shipping/storage cost due to lower volume of the product pre-assembly. A classic example

FIGURE 1.11 Adhesive bonding dispensing unit performing assembly on phone.

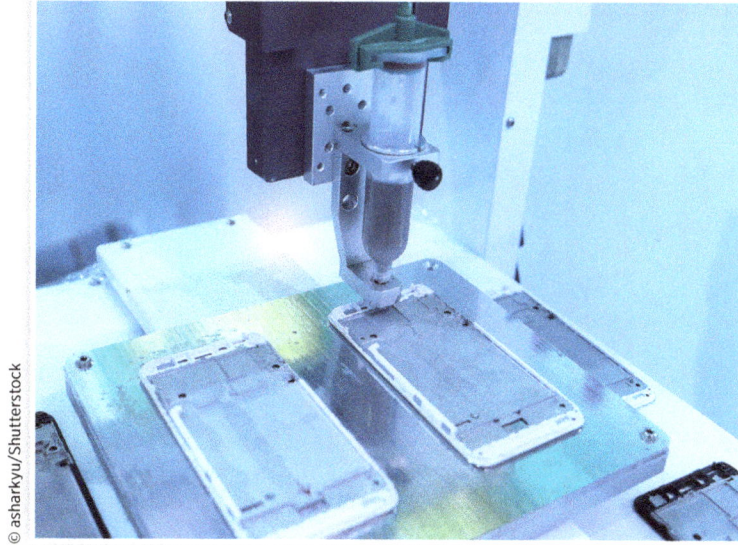

© asharkyu/Shutterstock

is IKEA with their DIY concept. A clear advantage for Mechanical Joining is the ability to disassemble the components without degradation of the components. This is desirable when we think of replacing a part that experiences faster wear than others, such as a car tire or battery. Welding processes are a cost-effective method to join components permanently. Although adopting welding as an assembly process is economically advantageous, it is challenging to perform disassembly. The components are bound to require rework. Bonding is the process of using adhesives to join two or more components. Bonding is widely adopted in industry for various applications (see Figure 1.11). It is worthwhile to highlight that composite bonding, detailed in Chapter 8, possesses multiple possibilities such as co-curing, co-bonding, and secondary bonding. Co-curing is the act of curing together two or more uncured components. Co-bonding is the act of curing one uncured component to another cured one. Secondary bonding is a standard adhesive bonding process of two fully cured composite parts.

1.2 Manufacturing Evolution

Manufacturing continuously defines the world. Once, countries were able to win wars based on their ability to mass produce (Manufacturing 2.0) military equipment. In our present time, mass production is not as important as innovative manufacturing. These different eras were recently categorized and labeled as

1. **Manufacturing 1.0/Industry 1.0**: It stands for the mechanization of machinery in contrast to manual processing. The transformation of energy sources to produce goods highlights the first era in manufacturing evolution (see Figure 1.12).

2. **Manufacturing 2.0/Industry 2.0**: It stands for the mass production of products. This has been particularly encouraged by the arming race between countries. It led mass production of airplanes, transportation vehicles in parallel to civil goods.

3. **Manufacturing 3.0/Industry 3.0**: It stands for the introduction of automation, robotics, and computers into manufacturing. CAM is one of the direct results of this era as well as the replacement of humans in hazardous working conditions by robotic arms.

FIGURE 1.12   Old steam machine: the mechanization of machinery is considered as the first stepping-stone in the evolution of manufacturing.

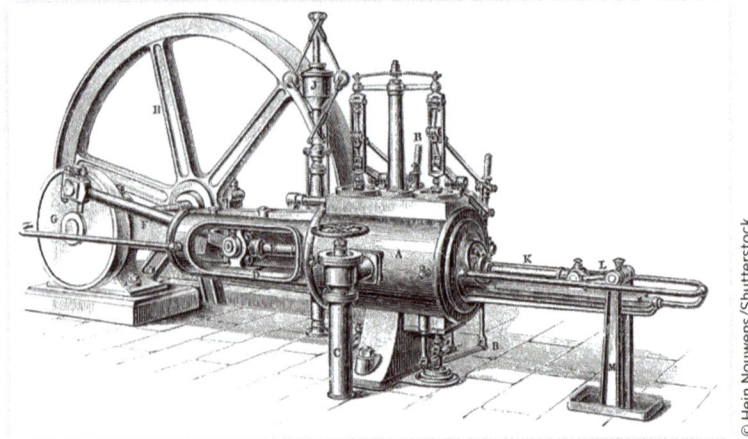

© Hein Nouwens/Shutterstock

4. **Manufacturing 4.0/Industry 4.0**: It stands for the integration of cyber-physical systems and the digital transformation of industries. The integration of cognitive manufacturing - smart systems that analyze and interprets data - is offering new insights and abilities for next-generation industries.

This classification, stemming from a German scientific project, is now widely adopted. In the United States, Industry 4.0 is often split into Smart Manufacturing, describing the analytics and data side involving all functions of a manufacturer, and Advanced Manufacturing, which focuses on innovations in processing technologies, novel materials, etc. Furthermore, a new term is currently trending, labeled Ubiquitous Manufacturing. The latter stands for the ability to "design anywhere, make anywhere, sell anywhere and at any time."

Chapter 9 develops Manufacturing Productivity fundamentals such as cycle time computation, cost estimation of the manufacturing process, and design optimization to facilitate manufacturing. In addition, Manufacturing Quality topics such as inspection methods, statistical tools, and design of experiments are also covered in this chapter. Manufacturing Productivity and Manufacturing Quality both became driving factors throughout the evolution of manufacturing from Industry 1.0 to Industry 4.0. Chapter 10 "Smart Manufacturing" introduces current developments of the digital transformation in manufacturing to augment the detailed description of Advanced Manufacturing topics covered throughout Chapters 2-8. We did not include a separate Design for Manufacturability (DfM) chapter in this book but chose to infuse the DfM principles throughout the entire book as we believe design and manufacturing are two sides of the same medallion that need to be deeply integrated and communicate.

The current evolution of manufacturing processes, manifested in the Industry 4.0 concepts and the digitization of manufacturing industries, is still an active research topic. The following sections present topics that represent some of the current manufacturing evolution topics of interest: Internet of Things (IoT), Digital Twin, and Cyber Manufacturing. Chapter 10 will provide more details on these topics.

1.2.1 (Industrial) Internet of Things (IIoT)

Nowadays, "things" are connected and transmitting data through the internet. The concept is dependent on (1) the capacity to collect and communicate the massive amount of data, (2) the ability to analyze, visualize, and communicate the data, and (3) the ability to make accurate conclusions/propositions (see Figure 1.13). The concept of IoT requires secure communication protocols and increases cybersecurity requirements stemming from more and more entities

FIGURE 1.13 Internet of things.

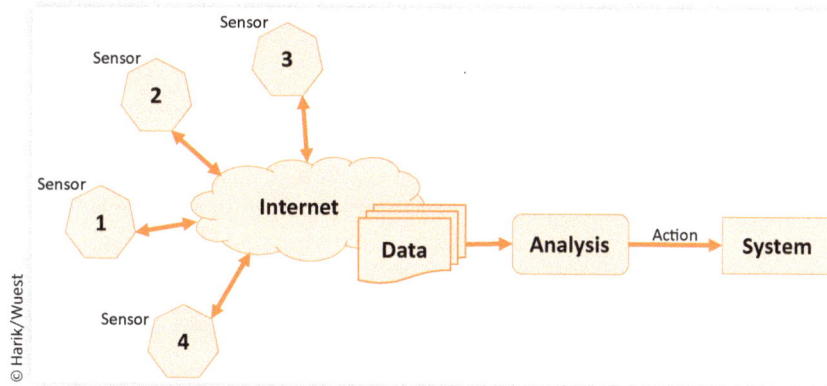

© Harik/Wuest

being connected to the network. The concept of IoT together with machine learning/AI provides a foundation for optimizing any process based on connected sensors and (real-time) data collection with little to no human intervention. Examples can be transportation systems, energy optimization, personal health, and manufacturing facilities. In an industrial setup, an IoT system, often referred to as Industrial Internet of Things (IIoT) can be monitoring multiple parameters, such as power and runtime, to draw inferences on expected malfunction or prospective downtime. This enables the industry to prepare for the asset maintenance ahead of failure in a proactive approach, often referred to as predictive maintenance. This development is gaining significant traction with the emergence of versatile Industrial Internet Platforms that are adopted by industry at a fast pace.

1.2.2 Digital Twin

The Digital Twin is an exact numerical replicate of the physical system that is connected and updated by sensor data (see Figure 1.14). The concept of the Digital Twin is to create an interactive blueprint of systems that enhance design, manufacturing, and servicing operations. The Digital Twin receives real-time operating conditions and runs predictive models to optimize asset task. Machine learning is a principal component of success for digital twins. Machine learning is the inherent capacity of the machine to learn from its operation and to adapt its performance. The machine will then, over time, be able to function without programming or direct human interference after learning from previous actions by the operator or similar systems in use. There are multiple machine learning approaches such as deep learning and genetic algorithms that can be used depending on the data available and circumstance of operation. Digital twins can be seen as an "Intranet of a Thing" where one unique system is sensed, its data collected and analyzed.

FIGURE 1.14 Digital twin of a physical system.

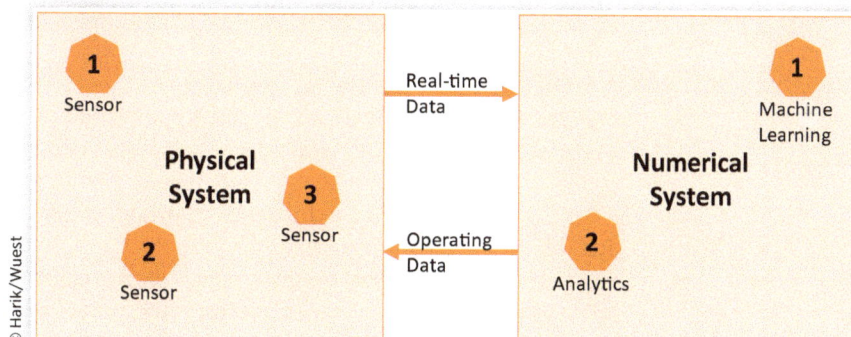

© Harik/Wuest

FIGURE 1.15 Classifications of engineering materials into metals, polymers, and ceramics.

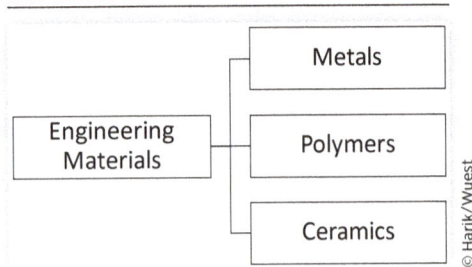

1.3 Manufacturing and Materials

1.3.1 Engineering Materials

Metals, Polymers, and Ceramics are the three natural categories of Engineering Materials (see Figure 1.15). Naturally occurring materials fall mostly also within one of these categories. Metals played a major role throughout history and are still essential for many applications. Iron and bronze defined two archaeological periods based on the discovery of tooling made out of these materials, the Bronze Age (~3300 BC to ~1200 BC) and Iron Age (~1200 BC to ~AD 800), respectively. The Earth's crust contains an abundance of metals such as aluminum, iron, and titanium. In addition, seawater contains these and other metals such as magnesium. In this case, the process to extract the pure metal from its $MgCl_2$ ore is however harder and more expensive. It was actually during a competition to replace the materials of pool table balls that scientists discovered polymers. The billiard balls were traditionally manufactured from ivory originated from elephant tusks. Polymers are generally less dense than most metals and ceramics with high corrosion resistance. Ceramics are compounds containing metallic and nonmetallic elements. The most famous ceramics are glass and carbides for tooling. While this section presents a detailed overview of metals, we present a more comprehensive overview of polymers in Chapter 7 "Polymers Manufacturing."

Design desired properties in general counter manufacturing desired properties. If a material is to have high strength, its machining process will be harder (and more expensive) in most cases due to, for example, the required harder cutting tools and the increased wear of those during operation. Obviously, material selection plays an important role in both design and manufacturing activities. However, although concepts like Design for Manufacturing and Design for Assembly exist, design and stress engineers often make material selections themselves without much support. Another equally important factor in the determination of the material and manufacturing selections is the availability of resources. Chapter 9 will discuss these concepts in detail.

We have intentionally omitted composite materials in this section. While composites are a premium engineering material, the separation is to highlight the difference between naturally occurring elements and composite elements. There exist natural occurring composites such as wood and human bones. However, the combination of two or more distinct phases of naturally occurring materials creates composites used in industrial applications. Section 8.1 of Chapter 8 details composite materials and their manufacturing.

1.3.1.1 **METALS**

Metals are isotropic materials that exhibit the same behavior along all of their directions. This is very important as it reduces the design complexity and indirectly the manufacturing requirements. Moreover, metals are highly abundant in nature and can be recycled comparatively well. Mining of selected metals is not easy, and/or their refinement processes are expensive. In such cases, recycling becomes one of the predominant forms of raw material production for such metals. Aluminum is an example where recycling is gaining traction since the energy requirements for recycling is less than 5% of what is needed for the refinement of aluminum from its naturally occurring ore.

Usually, a stand-alone category separates iron from other metals (see Figure 1.16). This is justified by the abundance of usage, superior properties, and relatively low cost. Iron-based metals are named Ferrous Metals. Non-Ferrous Metals include all other metals such as aluminum, titanium, magnesium, and others.

Table 1.1 lists the most common metals used in engineering applications. Superalloys and refractory and noble/precious metals are listed throughout the book where needed. Table 1.1 lists the principal ore extracted from natural resources.

FIGURE 1.16 Classification of metals.

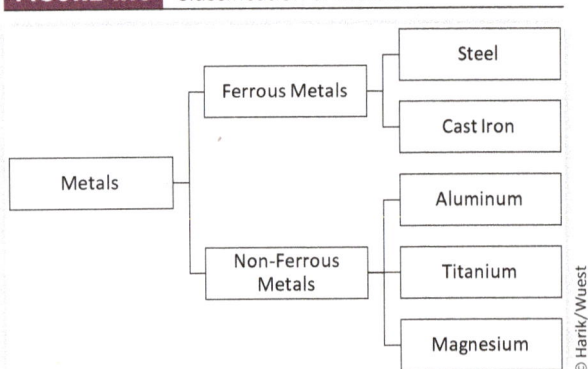

TABLE 1.1 Table listing most prominent metals and some of their properties

Name	Principal ore	Famous for	Application
Iron	Fe_2O_3	Most used metal	Above 70% world usage
Aluminum	Al_2O_3	Most abundant metal	Cans, automotive
Magnesium	$MgCl_2$	Lightest metal	Aerospace, missiles, bicycles
Copper	$CuFeS_2$	Oldest known metal	Electric wires
Nickel	$(Fe,Ni)_9S_8$	Iron copycat	Alloying to avoid corrosion
Titanium	TiO_2	Strength-to-weight ratio	Aircrafts, jet engines
Zinc	ZnS	Suitable for die casting	Alloy with copper (brass)
Lead	PbS	Heavy metal	Plumbing, x-ray shields
Tin	SnO_2	Suitable for solders	Alloy with copper (bronze)

© Harik/Wuest

Transformation of metals from their principal ores is typically through a crushing, washing, separation, and purification process. Multiple specific processes such as the Bayer Process, Electrolysis, Flotation, and Precipitations to extract the metal from their naturally occurring state. We also listed the principal application, not meant as an exhaustive list, and the reason each metal is famous.

Following we detail the iron-making process (see Figure 1.17) and the resulting Iron-based materials. Iron making requires a two-step process: (1) Iron reduction from hematite to Pig Iron, (2) Refinement of carbon and impurities concentrations from Pig Iron to steels and Cast Iron (see Figure 1.18). The first step intakes the naturally occurring Iron ore and breaks the molecule from Fe_2O_3 to Fe. Pig Iron results from this operation. Blast furnaces (see Figure 1.19) are a principal system used to perform this first step. It uses a dual solid/gas cycle: (1) Solids flows from the top (ores, coke, limestone) and are collected at the bottom (slag and Pig Iron), and (2) Gases flow from bottom up and support the melting, heating, and oxygen reduction process.

FIGURE 1.17 Steelworker when pouring liquid titanium slag from arc furnace.

© Shestakov Dmytro/Shutterstock

FIGURE 1.18 Iron making is typically made in two stages.

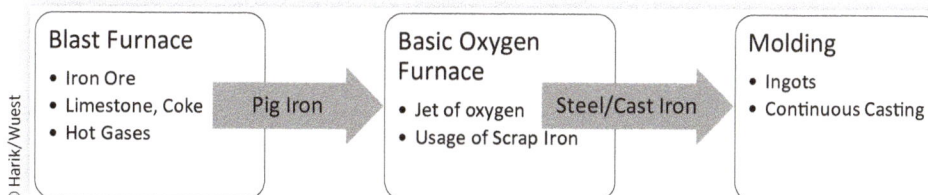

Blast Furnace
- Iron Ore
- Limestone, Coke
- Hot Gases

Pig Iron →

Basic Oxygen Furnace
- Jet of oxygen
- Usage of Scrap Iron

Steel/Cast Iron →

Molding
- Ingots
- Continuous Casting

© Harik/Wuest

FIGURE 1.19 Blast furnace cycles.

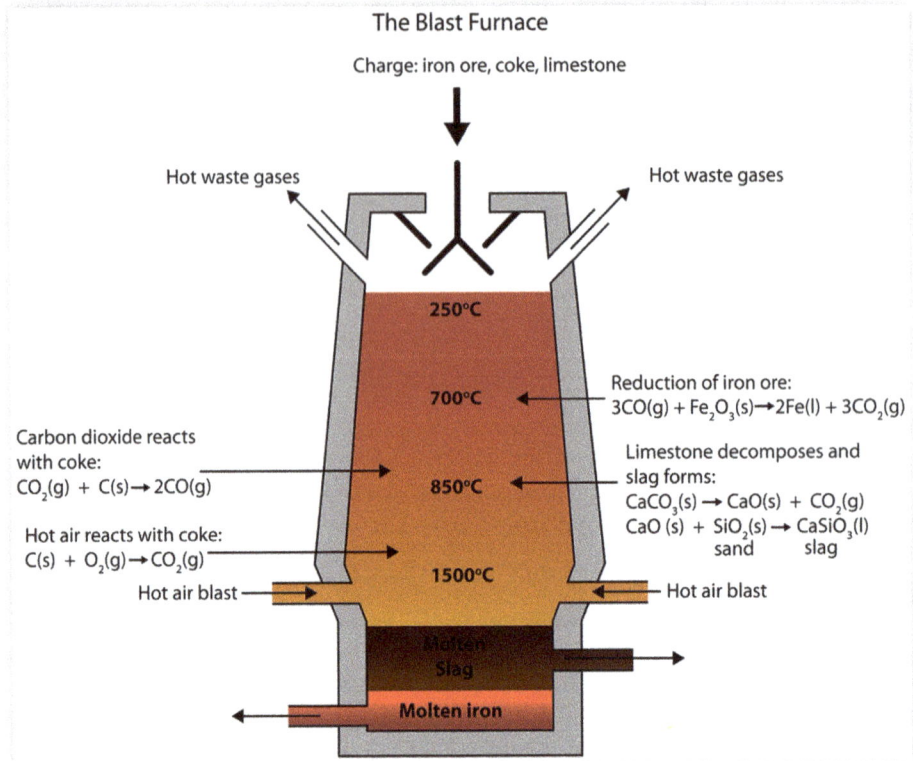

The Blast Furnace

Charge: iron ore, coke, limestone

Hot waste gases — Hot waste gases

250°C

700°C

Reduction of iron ore:
$3CO(g) + Fe_2O_3(s) \rightarrow 2Fe(l) + 3CO_2(g)$

Carbon dioxide reacts with coke:
$CO_2(g) + C(s) \rightarrow 2CO(g)$

850°C

Limestone decomposes and slag forms:
$CaCO_3(s) \rightarrow CaO(s) + CO_2(g)$
$CaO(s) + SiO_2(s) \rightarrow CaSiO_3(l)$
sand   slag

Hot air reacts with coke:
$C(s) + O_2(g) \rightarrow CO_2(g)$

Hot air blast

1500°C

Hot air blast

Molten Slag

Molten iron

© Steve Cymro/Shutterstock

The second step in iron making intakes Pig Iron and transforms it into steel or Cast Iron. Basic oxygen furnaces are a principal system used to perform this second step. It uses pure oxygen to burn off excess carbon content and other impurities. Pig Iron obtained from the first step contains 5-6% carbon. Although carbon increases the strength of Ferrous Metals, however it decreases its ductility and renders it brittle. We apply the refinement step until we obtain the desired carbon content.

Carbon content defines the different types of Ferrous Metals: (1) Cast Iron has concentrations of carbon higher than 2.1% and (2) Steel has concentrations of carbon less than 2.1%. Steel are designated using a SAE designation scheme (see Figure 1.20) of four digits: A, B, C, D. The first digit A grants the major classification: 1bcd for carbon steels, 2bcd for Nickel steels, and so on. The second digit B grants a secondary classification within the first category: 10cd for Plain Carbon and 72cd for Tungsten-Chromium steels with 1.75% by weight Tungsten and

FIGURE 1.20 SAE designation scheme.

A	B	CD
Major Classification	Secondary Classification	Carbon content % by weight
1	0	50
Plain Carbon		0.5% carbon by weight
3	4	20
Nickel-Chromium Steel	3% Nickel and 0.77% Chromium	0.2% carbon by weight

© Harik/Wuest

FIGURE 1.21 Polymers classification system for recycling purposes (SPI).

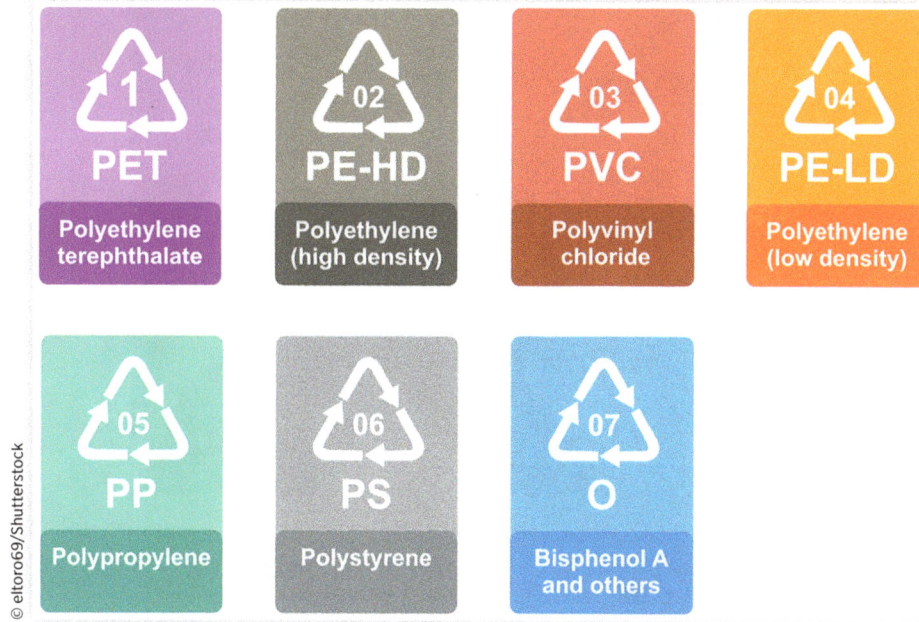

© eltoro69/Shutterstock

0.75% by weight Chromium. The last two digits CD represent the carbon content by weight: 1050 Steel is a plain carbon steel with 0.5% carbon content by weight.

1.3.1.2 POLYMERS

Polymers are chains of monomers grouped together. We identify three principal categories of polymers: Thermoplastics, Thermosets, and Elastomers (see Figure 1.21). Thermoplastics are commercially the most important variant of polymers and make up over 70% of the polymer market share. Thermoplastics can be recycled, however, experience material degradation to some extent. To facilitate their recycling the Society of the Plastics Industry (SPI) created visible labels (see Figure 1.21). When you look at the plastic products around you, such as your water bottle, you can find this stamp on many of them. Thermosets on the other side cannot be recycled and are as such not as desirable from a sustainability point of view. However, as they are the primary choice for certain applications, including as matrix material for composites, they cannot be avoided fully. Elastomers are part of the thermoset family and as such cannot be recycled in the regular way either. However, there are developments that try to combine the recyclability of thermoplastics and the material properties of thermosets, sometimes referred to as Thermoplastic Elastomers.

The way of linking the monomers influences the resulting polymer (see Figure 1.22). Polymers are a net shape process and are chemically processed. Polymer products range from plastic water bottles, PVC pipes, automotive components, tires, and many others. Lego bricks are also made out of thermoplastic materials. Chapter 5 introduces in detail the categories, properties, and fundamentals of polymers.

FIGURE 1.22 Polymers classification.

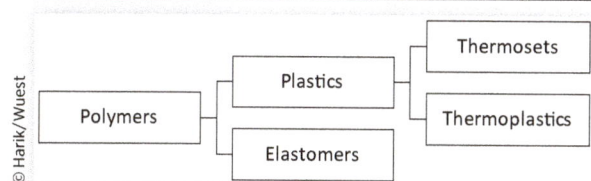

© Harik/Wuest

1.3.1.3 CERAMICS

Crystalline ceramics and glass constitute the two categories of Ceramics (see Figure 1.23). Crystalline ceramics include traditional ceramics such as pottery, and advanced ceramics such as Tungsten Carbide. Tungsten Carbide is a famous material used for machining cutting tools. Glass is a non-crystalline amorphous solid with widespread usage. Glass-Ceramics is different from glass. Glass-Ceramics have both a crystalline and an amorphous phase. They were discovered by mistakenly heating glass to higher

FIGURE 1.23 Ceramics classification.

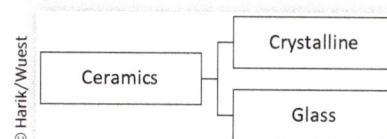

© Harik/Wuest

FIGURE 1.24 Bicycle carbon fiber wheel.

© taro911 Photographer/Shutterstock

than usual temperatures. This book does not specifically cover Ceramics Manufacturing as these are used mainly for specialty products or only in selected industries.

1.3.2 Composites

Composites are not to be confused with alloys, resulting from melting together two or more materials. When we join materials in their heterogeneous form, we obtain a composite material. Composites consist typically of two phases: Forming and Strengthening. The forming phase, labeled as matrix/resin, provides the formability and ductility property. We can thus manufacture complex geometries (see Figure 1.24). The strengthening phase, labeled as reinforcement/ fibers, provides the strength to the final part. Composites and composites manufacturing are expensive. Design of composites is also a complicated task. Composites possess unidirectional properties and are anisotropic. The design parameters are no longer strength, stiffness, and weight such as in metal design. Stacking sequence, fiber angles, distinct properties of the matrix and reinforcement stages, curing cycle and many other parameter selections are required. Section 6.1 of Chapter 6 details composite materials.

1.3.3 Properties of Materials

This section enumerates prominent physical and mechanical properties of materials (Table 1.2). Other textbooks typically present an extensive review of material properties including chemical, electrical, and magnetic properties. Our intention is to provide the best, and most up to date with current industry practice, coverage of manufacturing and its needs. We further elaborate properties where needed. That is, Chapter 2 includes details on the stress-strain relationship needed for Deformative Manufacturing.

1.4 Organization of the Book

This book contains 10 chapters. Chapter 1 presents a general introduction to manufacturing families, manufacturing evolution, and materials in manufacturing. Chapter 2 introduces Deformative Manufacturing including both forming and casting processes. Specifically, Chapter 7 introduces Deformative Manufacturing of polymers manufacturing such as extrusion, injection, and blow molding. Chapter 3 develops subtractive manufacturing topics such as milling, turning, and drilling. Chapter 4 initiates the reader to additive manufacturing main technologies such as fused filament fabrication, Selective Laser Sintering, and Stereolithography. Chapter 8 is tailored towards composites manufacturing, a subset of additive manufacturing

TABLE 1.2 Listing of principal material properties

Brittleness (opposite of plasticity)	Structure will shatter with little to no deformation under loading
Coefficient of thermal expansion	Structure will undergo modification in response to temperature variation
Corrosion resistance	Structure will resist degradation by oxidation
Ductility	Structure is capable of deformation under load
Durability	Structure responds better to degradation by functionality
Elasticity	Structure is capable of restoring original shape after load bearing
Electrical conductivity	Structure is capable of conducting electricity
Fracture toughness	Structure resists fracture despites enclosing crack
Hardness	Structure resists indentation
Melting point	Structure will change state from solid to liquid
Plasticity	Structure will absorb load and transform it into plastic deformation
Reactivity	Structure will release energy by itself
Specific heat	Heat capacity per unit mass of material
Stiffness	Structure will resist deformation under load
Specific stiffness	Stiffness/weight ratio
Young's modulus	Measure of stiffness
Surface energy	Interface energy at surface of structure
Surface roughness	Structure deviation from nominal shape
Tensile strength	Maximum tensile stress prior to necking/failure
Specific strength	Strength/weight ratio
Viscosity	Fluid resistance to gradual deformation
Yield strength	Separation point between elastic and plastic regions

© Harik/Wuest

processes. We introduce Automated Fiber Placement, a prominent composite manufacturing process, as part of Chapter 8. Chapter 5 details assembly processes such as mechanical joining and welding. Chapters 6, 9, and 10 present different manufacturing tools and recent trends. We introduce Computer-Aided Manufacturing in Chapter 6. Chapter 9 details manufacturing productivity and quality, including concepts like Cycle Time, cost estimation, design for manufacturing are communicated, and Inspection methods, statistical tools, and design of experiments. The book ends with an update on smart manufacturing.

Bibliography

National Association of Manufacturers (NAM), Retrieved from nam.org: http://www.nam.org/Newsroom/Top-20-Facts-About-Manufacturing/, July 25, 2018.

Wuest, T., *Identifying Product and Process State Drivers in Manufacturing Systems Using Supervised Machine Learning* (Cham: Springer, 2015), doi:10.1007/978-3-319-17611-6.

2

Deformative Manufacturing

Deformative Manufacturing represents processes where we transform a material from Shape A to Shape B without any addition or subtraction of material. With subtraction, we mean the intentional removal of material during the process leading to the reduction of volume, which is considered to be Subtractive Manufacturing (Chapter 3). However, it is important to mention that some Deformative Manufacturing processes produce excess material as part of the standard process. An example is the runner/riser/gate mechanisms during a casting process that are required to facilitate the manufacturing process. Ultimately, the fundamental concept of Deformative Manufacturing is that the volume of materials remains unchanged in the process. The change of material following thermal expansion/shrinkage that occurs during some deformative processes, e.g., casting processes, is not considered as changing the volume of the material in this case.

Deformative Manufacturing is one of the earliest manufacturing processes used by humans and was enabled by the discovery of new materials. Forging examples dates back as early as 4000 BC when copper was used to make weapons and tools in what we would now consider relatively primitive and simple shapes. The forging development is strongly correlated with casting processes and their development. Historically, copper was cast in Mesopotamia in 3200 BC to produce similar artifacts as the early forged shapes, while other sources go back even further stating 9000 BC as the earliest indication of casting processes. These processes, and several others such as rolling and extrusion, continue to evolve with the introduction of innovative and transformative technologies. With the introduction of the Bessemer steel-making process in the 1850s, the basis for an industrial scale production of high quality and economical steel was set. Enabling competitively priced and high availability steel triggered new applications of steel in many areas such as construction, transportation, and several others.

The structuring of manufacturing processes into different subsets and families is not consistent across manufacturing literature and textbooks. The classification that we present, where removing material is subtractive, adding materials is additive, and modifying the shape without changing the volume is deformative, stems from our desire to simplify and unify the classification process. In that regard, we join forming and casting processes under the Deformative Manufacturing classification.

We classify Deformative Manufacturing of metals according to the temperatures involved in the process: If the material is liquefied from its solid state to pouring temperatures T_p above its melting temperature T_m, we consider the process as casting/solidification. In all other cases,

FIGURE 2.1   Classification of deformative manufacturing.

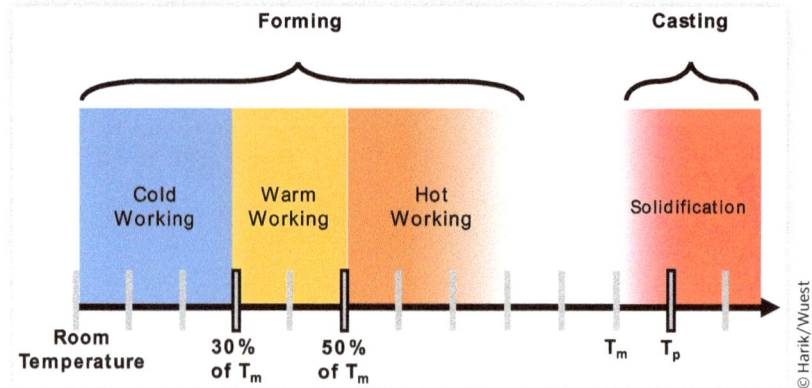

where the temperature ranges are below the melting temperature, we consider the processes as forming processes. Within forming processes, we can categorize three temperature classifications that can be applied to all forming processes: Cold Working, Warm Working, and Hot Working (see Figure 2.1).

Cold Working is the shaping of material at room temperature up to 30% of the melting temperature. It strengthens the metal through plastic deformation but renders it more brittle. Hot Working is the shaping of material above the recrystallization to 70/80% of the melting temperature depending on the material. This heating of the material allows using less force to transform the ingots. Finally, Warm Working is the shaping of material in between the Cold Working and Hot Working ranges. It offers a trade-off between Hot Working and Cold Working.

2.1 Understanding Material Behavior and Shape Factor

Material behavior when subjected to load is characterized by the stress-strain curve presented in Figure 2.2. Materials are typically subject to elastic, plastic, and necking regions. In elastic deformation, the shape undergoes a temporary change that is self-reversing once the load is removed. Structures follow Hooke's law in that region and have a linear relationship between stress and strain. The coefficient of this linear relationship is Young's modulus E which describes the stiffness of a solid material as a material property. Generally, a low Young's modulus describes a flexible material (shape changes significantly under elastic load), while a high Young's modulus is associated with a stiff material (shape changes marginally under elastic load).

FIGURE 2.2   Stress-strain curve.

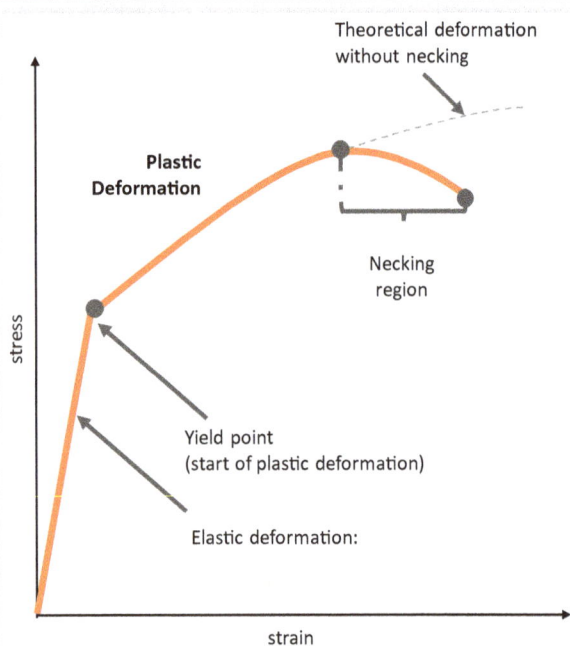

$$\sigma = Ee$$

$$e = \frac{L - L_0}{L_0} = \frac{h - h_0}{h_0} \qquad \text{Eq. (2.1)}$$

For deformative manufacturing to happen during the transformation process, permanent change of the shape is required and, as such, stress needs to reach the plastic deformation region beyond the yield point of the material. In that regard, the material is subjected to a stress value higher than the yield point. We communicate that manufacturing-enabling stress as *flow stress*. The latter is influenced by the strain rate, temperature, and other elements. It is imperative to know that the standard elongation formulation presented earlier cannot be used to determine the flow stress. The reason is that elongation uses a simplification where the elongation integration ratio denominator is replaced by the original length instead of the actual length. For manufacturing, since we need to be computing actual deformation values, we use *true strain ε* formulated as

$$\varepsilon = \ln\frac{L}{L_0}$$ Eq. (2.2)

where

   ε is the true strain
   L is the actual length
   L_0 is the original length

The computation of the *flow stress*, function of the strength coefficient, and the strain hardening exponent is

$$Y_f = K\varepsilon^n$$ Eq. (2.3)

To achieve a certain manufacturing operation, we often compute the *average flow stress* required to achieve the process. Average flow stress is computed as

$$\overline{Y_f} = \frac{K\varepsilon^n}{1+n}$$ Eq. (2.4)

PRACTICE PROBLEM

Determine the flow stress experienced during compression of a cylinder from a length of 50 mm to a length of 25 mm. The material has a strength coefficient of 200 MPa and a strain hardening of 0.2.

$$\varepsilon = \ln(L/L_0) = \ln(25/50) = -0.69$$ Eq. (2.5)

Negative value indicates compression instead of tension, BUT value is always understood as positive.

$$Y_f = K\varepsilon^n = 200(0.69)^{0.2} = 185.86 \text{ MPa}$$ Eq. (2.6)

What is the mean flow stress for the previous exercise?

$$\overline{Y_f} = (K\varepsilon^n)/(1+n)$$

$$\overline{Y_f} = 185.86/1.2$$

$$\overline{Y_f} = 154.88 \text{ MPa}$$ Eq. (2.7)

Another important consideration in Deformative Manufacturing is understanding the *shape factor*, its effects and the influence of different shapes on standard formulation. Typically, force is equal stress multiplied by the change area. However, let us consider the two shapes presented in Figure 2.3. If both shapes had the same area, is it logical that pushing them through a die would require the same force?

As such, the concept of the shape factor accounts for irregularity and nonuniformity of the change area. The introduction of a shape multiplier is used to correct the computation of the required force to achieve the desired manufacturing operation. The determination of the optimal shape factor is dependent on the specific manufacturing process at hand, such as forging or extrusion, and we present the different formulations in each section where applicable. Some of the shape factors are also dynamic in a sense that they change with the progression of the manufacturing process together with the change area. An example would be forging where the cross-sectional area undergoes modification as the manufacturing process is progressing.

We like to highlight that when a part goes through a Deformative Manufacturing process, the applied forces lead to a reduction of grain size and thus leads to change in the material

FIGURE 2.3 Shapes with different contour dimensions but same area.

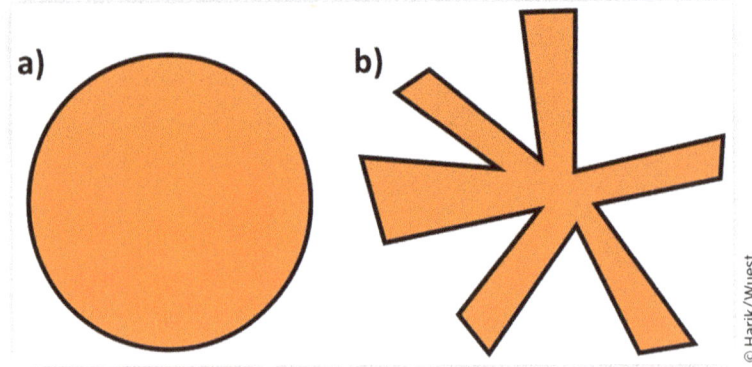

properties. This change is in most cases perceived as positive as it strengthens the part. This is the primary step in the design of microstructures. Lu et al. (2005) present a series of systematic experiments where they detail that the controlled introduction of nano-sized twins in pure copper leads to significant increase in the rate sensitivity of plastic flow.

2.2 Forging

Forging operations induce shape changes on the workpiece, which can be gradual or instantaneous, by plastic deformation under forces applied by various tools and dies. The applied forces can be hammering or pressing resulting in different mechanical properties. Forging processes are typically performed on bulk material. In contrast, sheet metals are often manufactured by sheet metal processes such as deep drawing and bending, as discussed later in this chapter.

A wide range of forging machines are available regarding their capacity to produce a wide range of parts from small to enormous, such as the Airbus 380 landing gear. Hydraulic presses can have up to 80k ton closed die pressure such as the hydraulic press owned by Shanghai Electric Group in China (Figure 2.4).

Forging can be categorized into (1) Open Die Forging, (2) Impression (Die) Forging, and (3) Upset Forging (Figure 2.5). The following sections will present each of these forging categories, followed by a section illustrating major defects that can occur during the forging process.

FIGURE 2.4 Forging machine.

FIGURE 2.5 Open forging processes.

Shen and Furrer (2000) state that advanced forging and heat treatment methods with enhanced material technology and computer modeling techniques are key elements for the success of modern forging processes used in the aerospace forging sector. The authors predicted further integration of process sensors, monitoring, and feedback systems to better adapt the process, readily available for ferrous-, titanium-, and nickel-based alloys – as well as other materials, and enhance it to better technical and commercial growth.

2.2.1 Open Die Forging

Open Die Forging is the simplest form of industrial forging where the height of a cylindrical billet (typically) is reduced. It does not require specialized tooling and has high flexibility. The process takes placed where the workpiece is deformed between two flat dies. In few instances, the dies are not flat but rather include some simple features and engravings. We typically use Open Die Forging when the workpiece is very large or when we have a small batch size. While the achievable shape complexity is limited within an open-die forging process, there are ways to increase the achievable shape complexity by strategically moving the part within the die during the process. An example for a creative and expert use of an open die process to create the desired form and function is the manufacturing of specialized aluminum alloy wheels for the "fastest car in the world" – the Bloodhound.

Figure 2.6 demonstrates the process of open die forging. In step 1, the material is loaded and the dies are closed. In the second step we are gradually applying a certain force value F_1. The shape keeps deforming as we progress with increasing the force. It is important to highlight that as the force is increased, the contact area between the die and the part is increasing. A side effect of most open-die forging processes is the occurrence of a flash, which is relevant for safety of operators, and the unavoidable bulging of the material in the die.

The strain in the open-die forging process is computed as

$$\varepsilon = \ln \frac{h_0}{h} \qquad \text{Eq. (2.8)}$$

The computation of the required force includes, along the flow stress and cross-sectional area, the shape factor for the open-die forging process:

$$F = Y_f K_f A_f \qquad \text{Eq. (2.9)}$$

FIGURE 2.6 Open die forging.

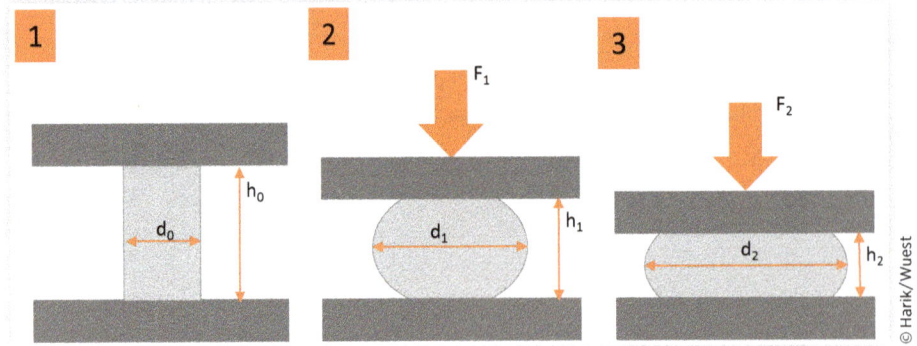

The shape factor for a part manufactured by an open-die forging process is computed as a function of the cross-section diameter, current instantaneous height, and coefficient of friction between the die and the part. We compute it as

$$K_f = 1 + \left(\frac{0.4\mu D}{h}\right)$$ Eq. (2.10)

Open die forging is often used as primary shaping process to pre-form the workpiece for subsequent processing such as closed-die forging, gradually changing the shape.

Han et al. (2018) discuss forging of mixed titanium composite that is reinforced, where open die forging was conducted. The study noted that the microstructures are inhomogeneous

PRACTICE PROBLEM

A material with the following properties (K = 300 MPa; n = 0.2) is open forged to half its height. The original shape is cylindrical having a height of 50 cm and a diameter of 40 cm. Given a coefficient of friction of 0.25 create a graph showing the incremental value of the forging force function of the strain.

Readers can download the excel template from http://introtomanufacturing.com. Under Resources (Figure 2.7)

FIGURE 2.7 Forcing forge function of strain.

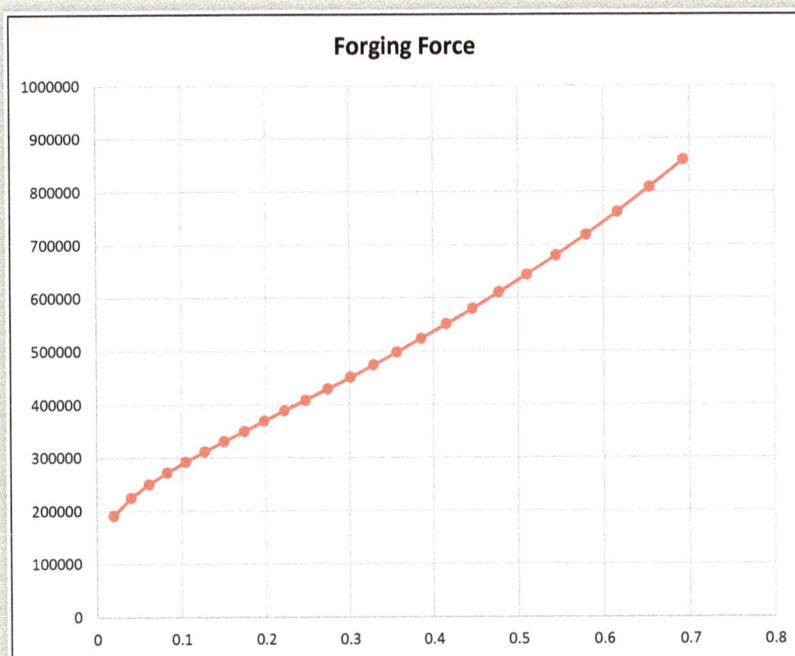

and that it can be mainly ascribed to the temperature variation in the open-die forging process. Tensile tests were conducted, and they confirmed that the center region exhibited higher strength than the peripheral region.

2.2.2 Impression Die Forging

Impression die forging is a rather complicated form of forging that includes some material loss in the location of the flash gutter (Figure 2.8). The principal concept is to obtain a near-net shape that typically needs minor finishing using subtractive manufacturing processes. The workpiece is first introduced in a principal stage between two dies that can have a complex shape. In a second stage, high pressure is applied on the bulky material to induce the permanent plastic deformation and material flow to fill the die cavities. Finally, in a final stage, the dies are completely sealed and the cavity between the dies is completely filled where the overflow of material is absorbed by the gutter location in a form of flash. This process enables also net shape processes, especially in parts that do not require tight tolerances in mating surfaces. An example of parts that can be produced without further processing is monetary coins. An example of parts that are produced in a near-net shape with need to subsequence material removal and finishing is engine blocks and piston heads.

Impression dies are expensive and can only be considered for large batch production to be economically feasible. The flexibility of forging in general is limited, and if the needed number of parts is low, resorting to additive or subtractive manufacturing is more advantageous and economically justified. An option is also to utilize expandable mold processes like sand casting for such small batch productions, in conjuncture with additive manufacturing processes. This way the "tooling" can be done using a simple additive manufacturing process, e.g., Fused Filament Fabrication to lower the cost and time needed. This specific combination of Deformative and Additive Manufacturing is referred to as Hybrid Manufacturing.

Gontarz et al. (2018) conducted experiments demonstrating that aircraft components made of magnesium alloy AZ61A can be produced by die forging on screw presses. The results of the mechanical and structural tests conducted in the study confirm that the products meet the required quality standards.

2.2.3 Upset Forging

Upset forging is most commonly used to manufacture fasteners with high throughput production (Figure 2.9). Based on output (number of produced parts) only, upset forging is the most used forging process globally. It is a deformative transformation of the shape through length compression and creation of a "tip." The sequence below details the process: First the wire raw material is fed towards the forging/die area. The punch then applies a force while the wire is maintained in its location leading to the formation of the tip. Both hot and cold upset forging are possible, and the process is applicable to workpieces ranging from small diameters to bars with 10 in. diameters. The standard design rule for upset forging is that the length of the deformed material is restricted to 3× its diameter. This prevents material buckling and, as such, failure of the manufacturing process.

FIGURE 2.8 Impression die forging.

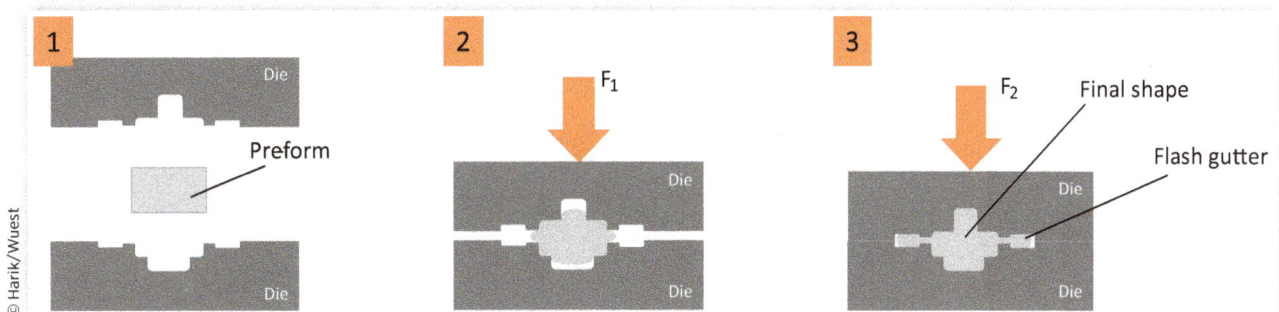

© Harik/Wuest

FIGURE 2.9 Upset forging.

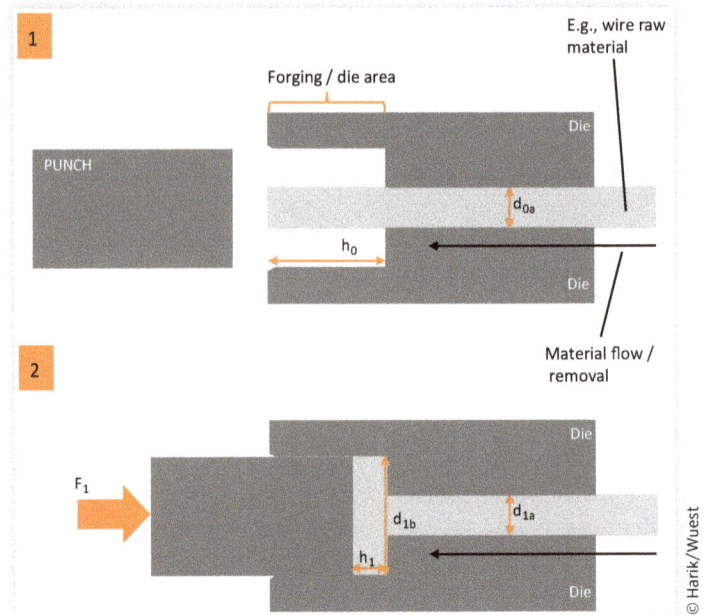

2.2.4 Common Defects in Forging

We list the prominent defects that can arise in forging applications (Figure 2.10):

- Incomplete die filling
- Die misalignment
- Forging laps
- Incomplete forging penetration
- (Material) Property variation (due to microstructural differences)
- Pitted surface (due to oxide scales, occurring at high temperature, sticking on die(s))
- Buckling (mainly common in upset forging due to high compressive stress)

FIGURE 2.10 Forging failure.

- Surface cracking (due to temperature differential between surface and center, and/or excessive working of surface at too-low temperature)

- Micro-cracking (due to residual stress)

2.3 **Extrusion**

Extrusion is a Deformative Manufacturing process during which the billet is reduced in cross section and/or formed in a difference constant cross-sectional shape by forcing the material to flow through a die orifice under high pressure. While extrusion is also used for Polymers Manufacturing, there is a distinct difference: metal extrusion is not a continuous process as the raw material is a billet with defined original length, while polymers extrusion can be continuously fed using pelleted plastics material. Certain shapes cannot be obtained through one single extrusion process, and as such multiple dies are sequentially placed to achieve the desired final cross section in this case. Most metals are typically hot extruded due to the significant forces required and are used to produce cylindrical bars, hollow tubes, or intermediate shapes that need subsequent manufacturing. Complex shapes can be extruded from softer metals such as aluminum. A typical example for an extruded part is a window frame section or profiles depicted in Figure 2.11.

Extrusion is often characterized by the reduction ratio of the cross-sectional area. This ratio is labeled *extrusion ratio* and computed as

$$r_x = \frac{A_0}{A_f}$$

Eq. (2.11)

The *true strain* computation is then

$$\varepsilon = \ln r_x = \ln \frac{A_0}{A_f}$$

Eq. (2.12)

However, to account for realistic extrusion friction, several models investigated a correction of the true strain and the computation of an extrusion strain accounting for these elements

FIGURE 2.11 Extruded aluminum profiles with constant cross section.

© D. Pimborough/Shutterstock

relevant to the material and the die angle. One of the most common adopted models is the Johnson model, where *extrusion strain* is computed as

$$\varepsilon_x = a + b \ln r_x \qquad \text{Eq. (2.13)}$$

where

ε_x is the extrusion strain
$a = {\sim}0.8$
$b = 1.2 - 1.5$ (if not provided, we assume that $b = 1.2$)
r_x is the extrusion ratio

2.3.1 Shape Factor

As stated earlier in this chapter, the shape complexity of the to-be-extruded part plays a major role in determining the required adjustment of force requirements. Standard extrusion equations were generated based on the reduction of a circular shape. The correction factor tries to compare the complexity of the shape, in contrast to a simple circular shape having the same area.

To determine the shape factor, we follow the four-step process highlighted below:

1. Calculate perimeter C_x of final shape

2. Calculate cross-sectional area A_x of final shape

3. Calculate perimeter of circle C_c with the same cross-sectional area as A_x

$$C_c = 2 * \sqrt{\pi * A_x} \qquad \text{Eq. (2.14)}$$

4. Computing Shape Factor K_x

$$K_x = 0.98 + 0.02 \left(\frac{C_x}{C_c} \right)^{2.25} \qquad \text{Eq. (2.15)}$$

It is worthwhile to note that any circular shape will have similar C_c and C_x, and, as such, always results in a shape factor of 1. This reflects that in this case, there is no shape-based increase in required force necessary to conduct the planned deformation.

PRACTICE PROBLEM

FIGURE 2.12

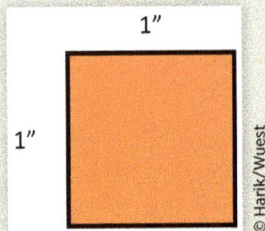

© Harik/Wuest

Find the shape factor of an extrusion die having a 1 in.² cross section (Figure 2.12).

Final shape perimeter

$$C_x = 4 \text{ in.} \qquad \text{Eq. (2.16)}$$

Final shape cross-sectional area

$$A_x = 1 \text{ in.}^2 \qquad \text{Eq. (2.17)}$$

Circle w/ A_x perimeter

$$C_c = 2 * (Pi * A_x)^5 = 3.545 \text{ in.} \qquad \text{Eq. (2.18)}$$

Computing Shape Factor

$$K_x = 0.98 + 0.02 (C_x / C_c)^{2.25}$$
$$= 0.98 + 0.02 (4/3.545)^{2.25} = 1.00624 \qquad \text{Eq. (2.19)}$$

2.3.2 Direct Extrusion

In direct extrusion, material is forced forward by a punch through the die orifice, producing a smaller cross section (Figure 2.13). It is imperative to remember that the volume of the material remains unchanged and the length of the final product increases, since the cross-section diameter is decreasing. This process is often referred to as forward extrusion.

The pressure needed for the extrusion process is reduced, as the process is ongoing. This is the computation of the pressure needed at the punch location:

$$p = K_x \overline{Y_f}\left(\varepsilon_x + \frac{2L}{D_0}\right) \qquad \text{Eq. (2.20)}$$

While most terms are straightforward and previously explained such as the shape factor, the average flow stress, and the extrusion strain, the final addition stems from the distance between the punch and the die location.

2.3.3 Indirect Extrusion

Indirect extrusion is formed around the punch by being forced backward around the punch within the die, producing hollow parts with solid bottoms (Figure 2.14). The operation is also characterized as backward extrusion and the bottom of the hollow parts should be thicker than the walls.

The pressure needed for the extrusion process is constant as the process is ongoing. This is the computation of the pressure needed at the punch/ram location:

$$p = K_x \overline{Y_f}\varepsilon_x \qquad \text{Eq. (2.21)}$$

The computation of the force at the punch is then the same for both direct and indirect extrusion:

$$F = pA_0 \qquad \text{Eq. (2.22)}$$

2.3.4 Combined Extrusion

Material is formed around the punch by being forced backward around the punch within the die and forced forward by the punch (Figure 2.15). Many designs include both extrusion forms; therefore, both forms are applied simultaneously to minimize production cost. There is a wide variety of different combinations possible depending on the creativity of the design and manufacturing engineer. As this process is very context specific, we refrain from providing generalized formulas in this case. To design and subsequently plan combined extrusion processes, high levels of expertise are required.

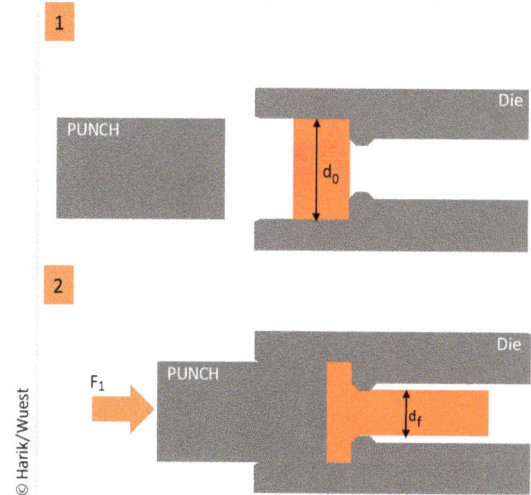

FIGURE 2.13 Direct extrusion.

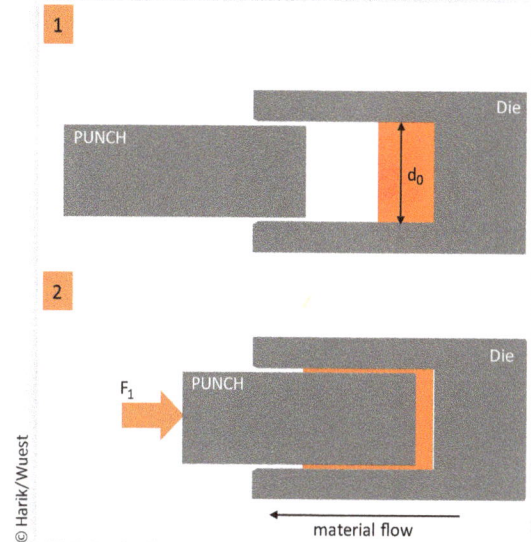

FIGURE 2.14 Indirect extrusion.

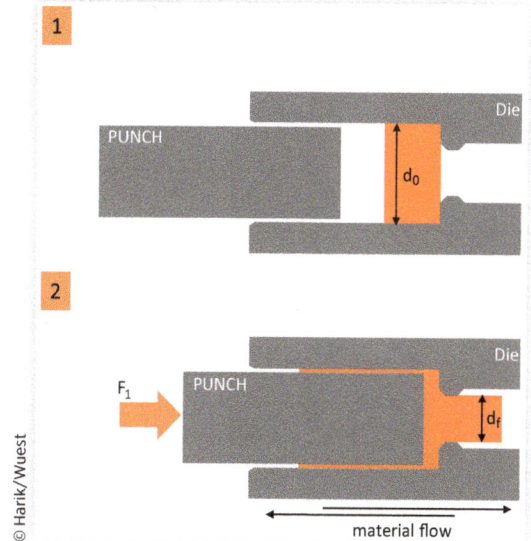

FIGURE 2.15 Combined extrusion.

CHAPTER 2

2.4 **Rolling**

Rolling is a deformation process where the part thickness is reduced by rolling it between one or multiple rolls. The volume of the original is maintained as expected in a Deformative Manufacturing process. Generally, rolling processes produce constant cross-sectional shapes, such as railroad rails. Typically, to achieve a certain cross section, a multi-stage process or tandem rolling is set to make it possible for the material to deform (Figure 2.16).

Rolling is a multi-stage process that can produce a large variety of different products. These products are typically split into sheet metal products, constant cross-sectional products, and miscellaneous products. Sheet metal products stemming from rolling processes include plates, sheets, large diameter pipes, etc. Constant cross-sectional products include seamless pipes, train rails, wires, bars, and various other profiles such as H, T, etc. Finally, miscellaneous products that are manufactured via rolling processes include rings, bolts, train wheels, etc.

Throughout the manufacturing textbook, we often use different terminology for the starting work material based on the manufacturing process and the geometrical shape. Billet, work part, bloom, slab, ingot, and many others refer to the starting part entering the manufacturing process. Bloom refers to square cross-sectional parts, slab to rectangular cross section, and billets are bars with cross-sectional area less than 6 in. × 6 in.

2.4.1 **Rolling Mills**

The most common setup of rolling mills is when two rolls rotating in opposite directions have a part intake with a certain thickness t_0 that is reduced to t_f (see Figure 2.17). Parts produced

FIGURE 2.16

© Usoltsev Kirill/Shutterstock

FIGURE 2.17 Rolling process.

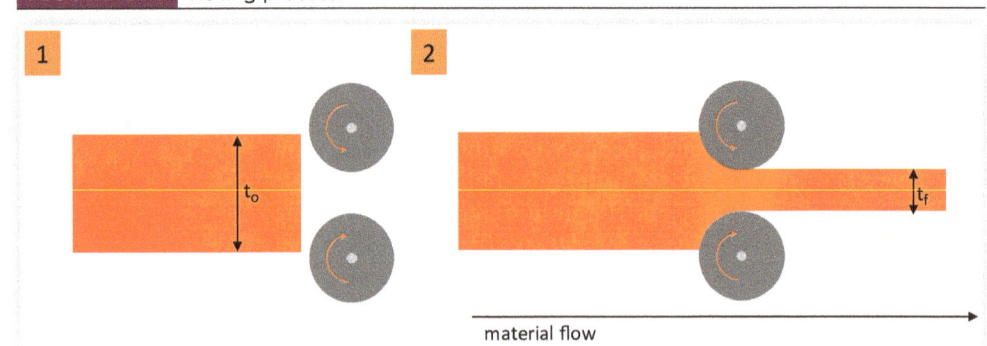

material flow

© Harik/Wuest

FIGURE 2.18 Schematic of rolling process sequence.

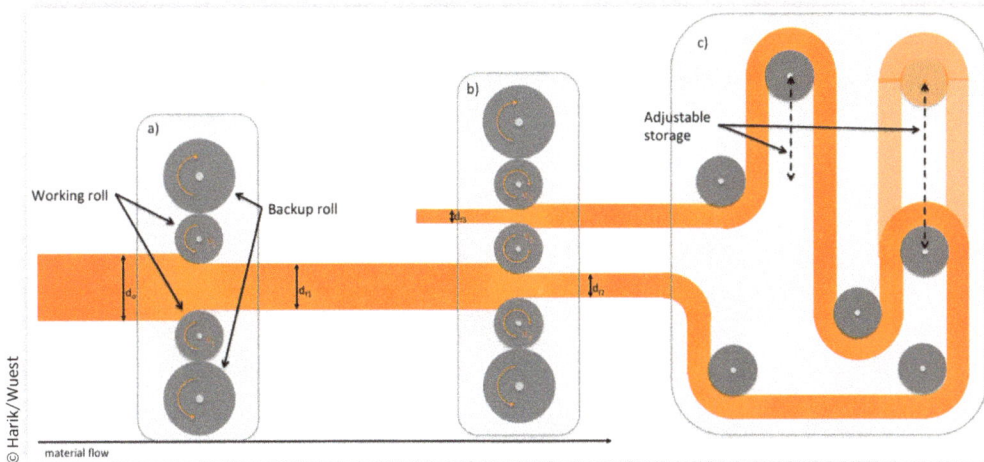

in this process are generally longer than those produced by extrusion, and larger than those produced by wire drawing.

This shaping of the material, by forcing it through the two rotating rolls, produces the newly reduced thickness part. There are many variations of different setups possible for rolling processes. Given the demanding environment comprised of, e.g., large forces and high temperatures, the rollers experience significant wear and stress, e.g., bearings and surfaces, throughout the process. Therefore, setups like the one depicted in Figure 2.18a are utilized to manage these challenges. In this arrangement, the backup rolls prevent the working rolls (that are in direct contact with the workpiece and experience the most wear and stress) from deflection and ensure the performance of the rolling system.

Similar to other manufacturing processes, in most cases a succession of subsequent rolling processes is required to achieve the required deformation. Rolling being a deformation process, which does not affect the volume of the part through the process. This leads to an increase in length of the workpiece during operation (given that width stays the same and height decreases) and in a multiprocess setup to different rotation velocities. This is illustrated in Figure 2.18, where the first and second set of rollers, while having the same diameter, have different velocities with $v_1 < v_2$ due to the increased speed of the material based on the length increase. While this is not a problem in a simple setup as depicted in Figure 2.17, it imposes an additional challenge when a more complex setup is used for the rolling operation. A schematic of such a more complex variation is illustrated in Figure 2.18.

In Figure 2.18b, the middle roll serves basically as an essential part of two deformation processes: deforming the material from d_{f1} to d_{f2} and simultaneously from d_{f2} to d_{f3}, rolling the workpiece in a forward and return pass. In such a setup, one of the main advantages is the reduced motor power and transmission requirement of the overall system. However, as the length of the material changes when the height is reduced, the material would feed into the second (return) pass at a different speed (higher) than in the forward pass. As the roll can only operate at one speed, an adjustable storage setup is needed that synchronizes the material feed (speed) for the process accordingly. There are different approaches how to synchronize the material feed, adjustable rollers as detailed in Figure 2.18c is one commonly used way. An industrial example of such a system is presented in Figure 2.19.

Rolling produces (continuous) parts with constant cross-sectional areas as mentioned before. While straight (surface) rollers are common in rolling operations, there is a wide variety of different shapes of rolling surfaces possible. One example is rollers producing train tracks with their typical T-shaped form.

The draft for flat rolling is defined as the difference between the starting thickness and the final thickness, and computed as

$$d = t_0 - t_f$$ Eq. (2.23)

FIGURE 2.19 Adjustable storage.

We define rolling reduction as

$$r = \frac{d}{t_0}$$

Eq. (2.24)

Considering the conservation of volume, one can write the following equation equating between the starting material and the final material:

$$t_0 w_0 L_0 = t_f w_f L_f$$

Eq. (2.25)

Deriving the conservation of volume equality and accounting for time, we can estimate the exit and entry velocity relationship as

$$t_0 w_0 V_0 = t_f w_f V_f$$

Eq. (2.26)

There exists a relationship between the Roll radius R, the coefficient of friction between the rolls and the materials, and the maximum possible thickness reduction/draft in a rolling operation. This relationship is as follows:

$$d_{max} = \mu^2 R$$

Eq. (2.27)

2.4.2 Thread Rolling

Thread rolling deforms the material by rolling it between two threaded dies to imprint external threading on a round bar/wire material. The threaded dies can be plane or round depending on the process, and thread rolling is a cold forming process that produces external threads with properties like strength, precision, uniformity, and smoothness.

Figure 2.20 depicts one common variation of thread rolling dies that uses a flat surface to imprint the structure on the preformed workpiece, e.g., a bolt. Another variation is to use two round dies to impregnate the threads on the workpiece. An industrial setup of such a thread rolling die is illustrated in Figure 2.21.

FIGURE 2.20   Schematic of thread rolling using a flat surface die.

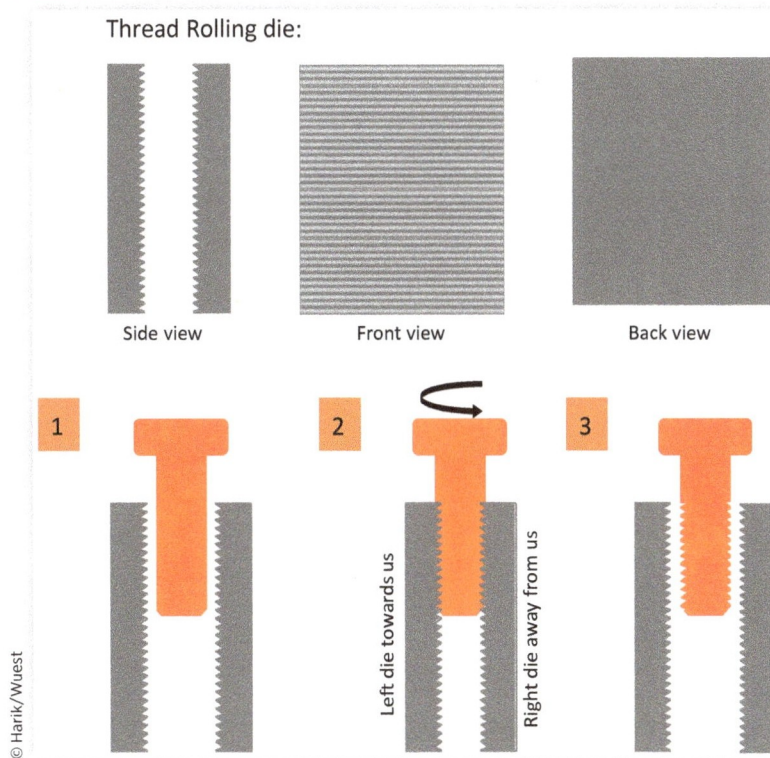

Thread Rolling die:

Side view   Front view   Back view

1   2   3

Left die towards us

Right die away from us

© Harik/Wuest

FIGURE 2.21   Industrial thread rolling tooling using round dies.

© Matee Nuserm/Shutterstock

2.4.3 Ring Rolling

Ring rolling is a deformation process that increases diameters of rings by reducing the radial thickness of the workpiece between a set of rollers (Figure 2.22). Workpiece volume is maintained which leads to diameter increase and requires rollers to adjust with workpiece throughout the ring rolling process, following a similar principle to the adjustable storage in a stacked rolling setup presented before. The driver roll is fixed in most cases while idle roll, rolling pins, and guiding rolls adjust to the changing dimensions of the workpiece. The reason the driver roll is commonly fixed is that the complexity of connecting transmission and powertrain increases significantly when the part is movable. Having the non-powered rolls adjustable allows

© Harik/Wuest

FIGURE ? ??   Schematic of ring rolling

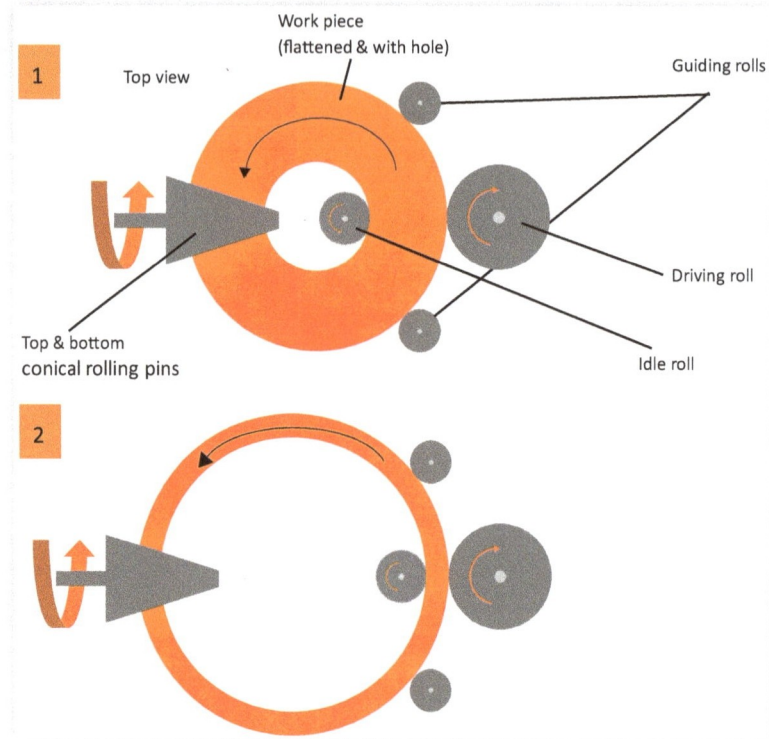

for the system to function while keeping complexity and maintenance manageable. In addition to the driving roll, idle roll, rolling pins, and guiding rolls, two, often conical-shaped, rollers maintain the upper and lower boundary of the workpiece.

Ring rolling is generally a hot working process and results in desirable circumferential grain structure achieving better mechanical properties.

2.5 Casting

Casting is a prehistoric technology enabled by the use of fire. Early molds were made of stone, using stone carving to prepare the tooling (Figure 2.23). It was not until the sixteenth century that sand was used as mold material (in France). According to the ASM Handbook, the earliest metal objects of wrought native copper dates back to 9000 BC. Casting is a primary manufacturing process that transforms the shape of a material by changing its state to allow it to fill a mold cavity and take the shape represented without changing its volume. Although predominantly applied on metals, casting is applicable to all engineering materials, including polymers and ceramics. In this section, we will focus mainly on metal casting. Polymers processing using molds, such as injection and blow molding, are presented in detail in Chapter 7 of this book. While casting is suitable for mass production and capable of creating complex components in a versatile process, parts are often limited in mechanical strength and require subsequent manufacturing steps. Moreover, ergonomics of the casting process is challenging, as heat hazards are omnipresent in casting environments.

FIGURE 2.23   Sand casting.

© Marzolino/Shutterstock

FIGURE 2.24 Standard sand casting mold.

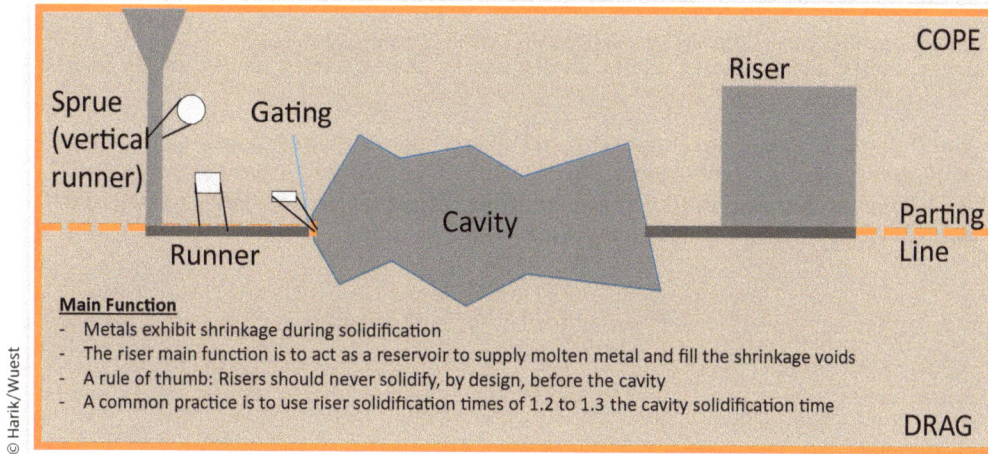

Sprue (vertical runner)

Gating

COPE

Riser

Cavity

Parting Line

Runner

Main Function
- Metals exhibit shrinkage during solidification
- The riser main function is to act as a reservoir to supply molten metal and fill the shrinkage voids
- A rule of thumb: Risers should never solidify, by design, before the cavity
- A common practice is to use riser solidification times of 1.2 to 1.3 the cavity solidification time

DRAG

© Harik/Wuest

Casting can be separated into two families. If the mold material is a solid material (e.g., stone, metal, ceramic) and is reused in the casting process to produce multiple subsequent parts, we refer to permanent mold processes. If the mold material is sacrificed to obtain the final shape of the workpiece (e.g., sand, wax), and we have to create a new mold every time we manufacture a new part (one part per mold), we refer to expendable mold processes.

The schematic in Figure 2.24 shows a standard sand casting mold, an expandable mold process, with its different constituent. The below schematic is simplified as in reality: multiple internal cores are needed to create functional parts. For example, a car engine block mold is made of at least 15+ different cores and molds.

The fundamental concept of casting is by (1) melting the material into a highly plastic (polymers) or liquid (metals) state, (2) designing a mold (and cores if applicable) to contain the molten material, (3) pouring the material into the designed mold, and finally (4) waiting for the material to solidify/cool.

When planning a casting process as depicted in Figure 2.25, we need to be able to answer multiple questions. Some of the main ones are listed here:

- What is the amount of heat needed to melt the metal up to its pouring temperature?

- How do we design the mold in the most advantageous way to enable smooth pattern removal (if needed)?

FIGURE 2.25 Casting process.

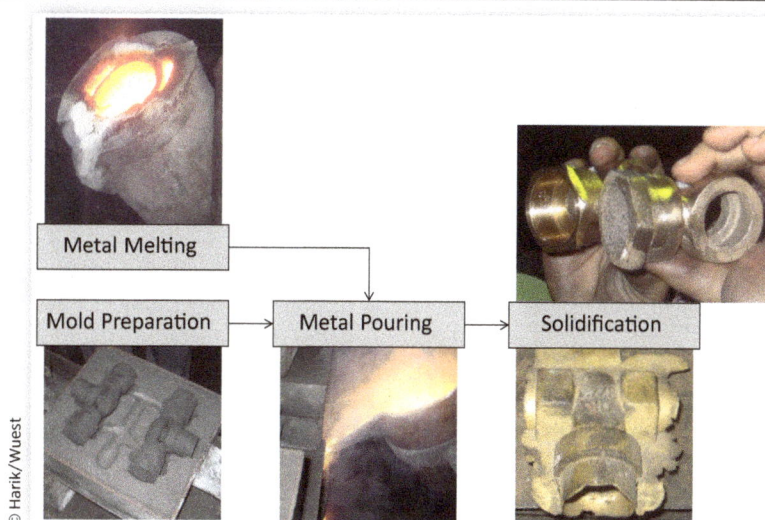

Metal Melting

Mold Preparation → Metal Pouring → Solidification

© Harik/Wuest

- How do we regulate the flow, and what is the pouring velocity/runner system design configuration?
- How long does the part need to solidify and to return to its solid state?

To answer the above questions, we can use the following equations:

- The amount of heat needed to get the metal up to its pouring temperature is equal to the sums of heat needed to get the metal to its melting temperature, the heat of fusion, and the heat to get the metal to 100°C above its melting temperature. The latter is referred to as pouring temperature. This is computed by

$$H = \rho V \left[C_s \left(T_m - T_0 \right) + H_f + C_L \left(T_p - T_m \right) \right]$$ Eq. (2.28)

where
    H is the total required heat
    ρ is the density
    V is the volume of to be heated metal
    C_s is the weight specific heat for solid material
    T_m is the metal melting temp.
    T_0 is the starting temp. – surrounding temp.
    H_f is the heat of fusion
    C_L is the weight specific heat for liquid material
    T_p is the pouring temp.

- To facilitate the design of the mold, following thoroughly the Design for Manufacturing and Assembly (DFMA) guidelines are required.
- With respect to the regulation of flow, an appropriate design of the pouring cup and sprue will ensure a laminar flow that will not erode the walls of the mold. This is computed by

$$v = \sqrt{2gh}$$ Eq. (2.29)

where h is equal to the height of the sprue.

- Once we have the value of the velocity flow, we can calculate the volumetric flow Q filling the mold, and the time needed to fill the mold. This is computed by

$$Q = vA$$ Eq. (2.30)

$$T_{MF} = \frac{V}{Q}$$ Eq. (2.31)

- Finally, the solidification time needed for the part to cool down is computed according to Chvorinov's rule that is dependent on the mold material, the (part) material, and the volume/area ratio of the mold cavity. This is computed by

$$T_{ST} = C_m \left(\frac{V}{A} \right)^n$$ Eq. (2.32)

where
    T_{ST} is the solidification time
    V is the volume of casting
    A is the surface area
    C_m is the mold constant
    n is mold exponent, usually = 2 for calculations

SAMPLE PROBLEM

What is the solidification time of a cube having a 6 cm side if the mold constant is 1 min/cm²?

Volume of the cube is $a^3 = 6^3 = 216$ cm³

Contact area of the cube is $6 * a^2 = 216$ cm²

$$t_{ST} = C_m (V/A)^n = 1 \text{ min} \qquad \text{Eq. (2.33)}$$

Design a cylindrical riser to solidify 30 s after the cube. The height of the cylinder is equal to its diameter.

Volume of the cylinder is

$$V = Pi * (d/2)^2 * d = Pi * d^3/4 \qquad \text{Eq. (2.34)}$$

Contact area of the cylinder is

$$A = 2 * \left(Pi * (d/2)^2\right) + 2 * Pi * (d/2) * d$$
$$= 2 * Pi\left(d^2/4 + 2 * d^2/4\right) = Pi * 3 * d^2/2 \qquad \text{Eq. (2.35)}$$

$$t_{ST} = C_m (V/A)^n = 1.5 \text{ min} \qquad \text{Eq. (2.36)}$$

$$1.5 \text{ min} = 1 \text{min/cm}^2 \left[\left(Pi * d^3/4\right) / \left(Pi * 3 * d^2/2\right)\right]^n$$
$$= [d/6]^2 \qquad \text{Eq. (2.37)}$$

$$d = 7.35 \text{ cm}$$

PRACTICE PROBLEM 2

What is the relationship between the diameter of a sphere D and the length of a cube side a if they have the solidification time of the sphere is half the one of the cube (Figure 2.26)?

FIGURE 2.26

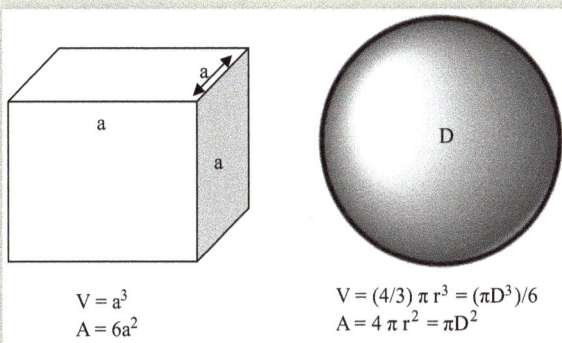

$V = a^3$
$A = 6a^2$

$V = (4/3) \pi r^3 = (\pi D^3)/6$
$A = 4 \pi r^2 = \pi D^2$

© Harik/Wuest

$$\left(\frac{V}{A}\right)_{sphere} = \frac{\frac{\pi D^3}{6}}{\pi D^2} = \frac{D}{6} \qquad \text{Eq. (2.38)}$$

$$\left(\frac{V}{A}\right)_{cube} = \frac{a^3}{6a^2} = \frac{a}{6} \qquad \text{Eq. (2.39)}$$

$$T_{solidification} = C_m \left(\frac{V}{A}\right)^2 \qquad \text{Eq. (2.40)}$$

$$T_{sphere} = \frac{T_{cube}}{2} \qquad \text{Eq. (2.41)}$$

$$2\left[C_m \left(\frac{V}{A}\right)^2\right]_{sphere} = \left[C_m \left(\frac{V}{A}\right)^2\right]_{cube} \qquad \text{Eq. (2.42)}$$

$$\left(\frac{V}{A}\right)_{sphere} = \frac{1}{\sqrt{2}}\left(\frac{V}{A}\right)_{cube} \qquad \text{Eq. (2.43)}$$

$$\frac{D}{a} = \frac{1}{\sqrt{2}} \qquad \text{Eq. (2.44)}$$

2.6 Sheet Metal

Sheet metal deformation describes the change of shape of a material in sheet form with no or little change in thickness. Generally, the deformation is two dimensional, and the modulus value of the material does not change. The elastic recovery of the material has to be accounted for, as it is usually larger than in other deformation processes (Figure 2.27).

FIGURE 2.27 Man working with sheet metal and special machine tools for bending.

FIGURE 2.27 Man working with sheet metal and special machine tools for bending.

© Shutterstock

2.6.1 Blanking

The sheet metal blanking process describes the process of removing the workpiece from the blank by forcing a shaped punch through the sheet into a shaped die (Figure 2.28). The remaining portion in the original sheet metal is then scrapped and used in recycling. Blanking and punching are often sequential in sheet metal manufacturing.

2.6.2 Punching

The sheet metal punching process describes the process of removing material from the blank which is the desired workpiece. The removed material is scrapped and reused in recycling. Punching is typically performed prior to blanking if needed for certain parts that have hollow areas in them (Figure 2.29).

FIGURE 2.28 Blanking.

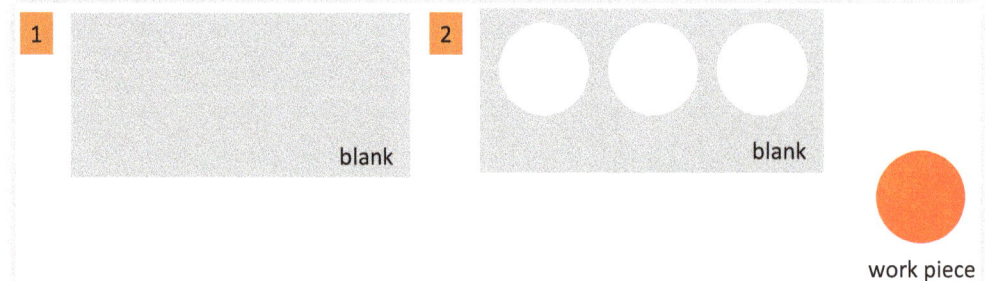

© Harik/Wuest

FIGURE 2.29 Punching.

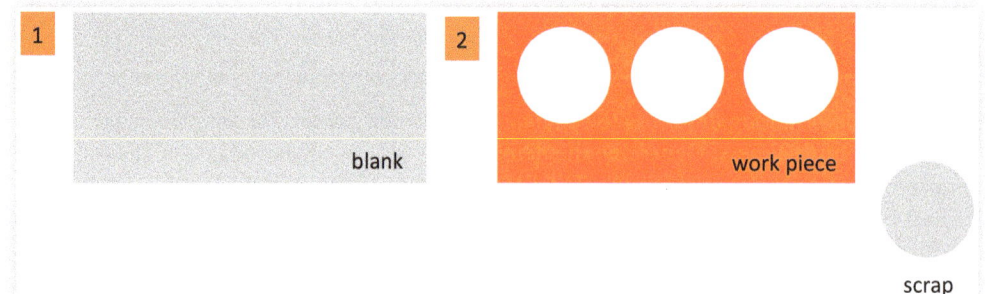

© Harik/Wuest

FIGURE 2.30 Bending process.

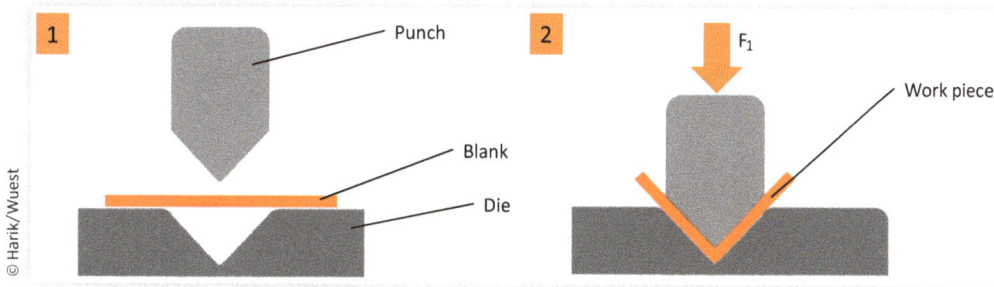

FIGURE 2.31 Drawing process.

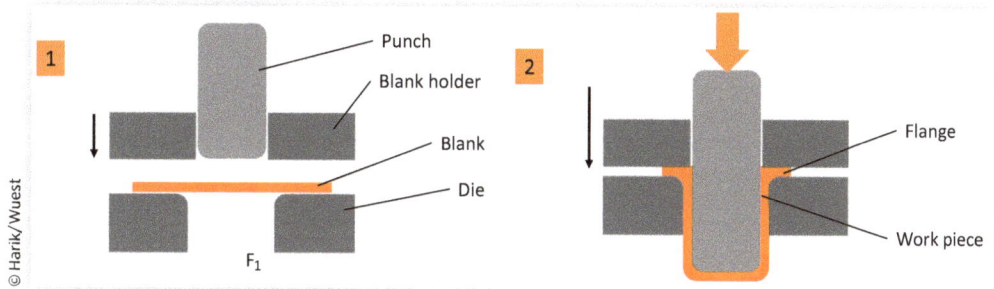

2.6.3 Bending

Sheet metal bending processes are plastic deformation processes around a linear axis while not (or only slightly) changing the surface area. Bending processes entail a certain elastic recovery from the combined tension and compression that needs to be accounted for. Parameters influencing the elastic recovery include material type and thickness. It is very important to highlight that the elastic recovery is typically manifested in a springback effect. Springback is when the bending angle is slightly increased by 2°-4° following the bending process (Figure 2.30).

2.6.4 Drawing

Sheet metal (deep) drawing describes a process of forming a material through plastic deformation using a punch and die, typically creating cylindrical or rectangular containers. There are two common types of sheet metal drawing processes, deep drawing (depth > diameter) and shallow drawing (depth < diameter).

References

Gontarz, A., Drozdowski, K., Dziubinska, A., and Winiarski, G., "A Study of a New Screw Press Forging Process for Producing Aircraft Drop Forgings Made of Magnesium Alloy AZ61A," *Aircraft Engineering and Aerospace Technology* 90, no. 3 (2018): 559-565.

Han, J., Lü, Z., Zhang, C., Zhang, S. et al., "The Microstructural Characterization and Mechanical Properties of 5 vol.%(TiBw+ TiCp)/Ti Composite Produced by Open-Die Forging," *Metals* 8, no. 7 (2018): 485.

Lu, L., Schwaiger, R., Shan, Z.W., Dao, M. et al., "Nano-Sized Twins Induce High Rate Sensitivity of Flow Stress in Pure Copper," *Acta Materialia* 53, no. 7 (2005): 2169-2179.

Shen, G. and Furrer, D., "Manufacturing of Aerospace Forgings," *Journal of Materials Processing Technology* 98, no. 2 (2000): 189-195.

3

Subtractive Manufacturing

Subtractive Manufacturing represents processes where we transform a part from Shape A to Shape B by subtraction of material. The fundamental concept is that we reduce the volume of materials throughout the process. Material is removed in a wide variety of ways. We commonly structure Subtractive Manufacturing processes in three categories: Conventional, Abrasive, and Non-traditional. Table 3.1 provides a high-level, non-comprehensive overview of selected Subtractive Manufacturing processes and their association with these categories.

This chapter will detail the most common traditional Subtractive Manufacturing processes, often referred to and summarized as machining in the scientific literature: *Milling, Drilling, and Turning*. These processes have in common, in contrast to abrasive and non-traditional subtractive processes, geometrically defined cutting edges. For these types of processes, we know the number and geometry of the cutting edge that attacks the material. Our coverage approach is a quantitative one that enables the reader to connect CNC machine parameters with the machining conditions. A necessary precursor to this chapter is the presentation of Shape Classification and Process Planning. Shape Classification enables the reader to connect, in the context of Subtractive Manufacturing, the to-be-manufactured shape defined by the engineering design with the appropriate manufacturing classification. Process Planning lists the functions required to appropriately analyze the part and define the subsequent cutting conditions (parameters) of the Subtractive Manufacturing process. The selection of optimal cutting conditions depends heavily on the agreed-upon manufacturing outcome objective, and can range, e.g., from low-cost to best achievable quality, or a compromise of these extremes. We conclude this chapter with a case study on manufacturing plans. Subtractive Manufacturing is pursued for its good dimensional accuracy and wide availability (resources and experience) in industry, whereas its principal disadvantage is material waste, mainly in the form of chips.

TABLE 3.1 Overview of Subtractive Manufacturing categories and processes (selected)

Conventional Subtractive Manufacturing Processes (machining)	Abrasive Subtractive Manufacturing Processes	Non-traditional Subtractive Manufacturing Processes
Milling	Grinding	Chemical: • Etching • Photochemical machining
Drilling	Honing	Mechanical: • (Abrasive) water-jet cutting • Ultrasonic machining
Turning	Lapping	Thermal (electric or chemical): • Laser beam machining • Plasma arc machining • Electron beam machining • Wire electrical discharge machining (wire EDM)
Shaping	Vibratory finishing	
Broaching		
Sawing		

© Harik/Wuest

3.1 **Shape Classification**

Selecting which Subtractive Manufacturing process is most suitable to create the physical manifestation of the designed part depends strongly on the geometry of the latter. By default, the process planner is motivated to select a combination of (1) the least complicated (aka least expensive in most cases) manufacturing process that can achieve the required shape, tolerances, and surface quality, as well as (2) the manufacturing resources available locally at the workshop, in close proximity and/or at a trusted partner's location. Process Planning is in essence the match-making process between design requirements and available manufacturing resources. Eventually, the process planner selects the machinery, fixtures, tools, cutting conditions, and manufacturing operations to achieve the desired manufacturing outcome.

A first, high-level classification of geometric parts can be (1) prismatic parts and (2) revolution parts (see Figure 3.1). *Prismatic parts* are extruded from a profile along an extrusion orientation. They are labeled 3-axis parts. When we further manipulate the part and include other features, such as pockets and multi-axis pockets and contouring, we obtain 5-axis parts. Both 3-axis and 5-axis parts are manufactured on a milling machine. *Revolution parts* are revolved from a profile along a revolution axis. A lathe machine is generally used to generate revolution parts using a turning process. All parts can include axial features, such as drilling holes (Figure 3.2). While Drilling is technically a milling process, it is so common in industry and our daily lives that we have a dedicated drilling machine (aka 1-axis milling) identified as drill press.

Finally, some parts can include both prismatic and revolution features. Those are manufactured either on a mill-turn machine or through a manufacturing plan that includes (at least) two phases: one on a lathe and another on a mill. This setup is rather common as turning is generally less complicated and the machine tools less expensive than high-end milling machines. Therefore, if we can generate a larger portion of the transformation on a lathe, with only certain features being induced by a 5-axis mill, this is often the preferred plan of action. In this case,

FIGURE 3.1 (a) Prismatic parts (milling), (b) revolution parts (turning).

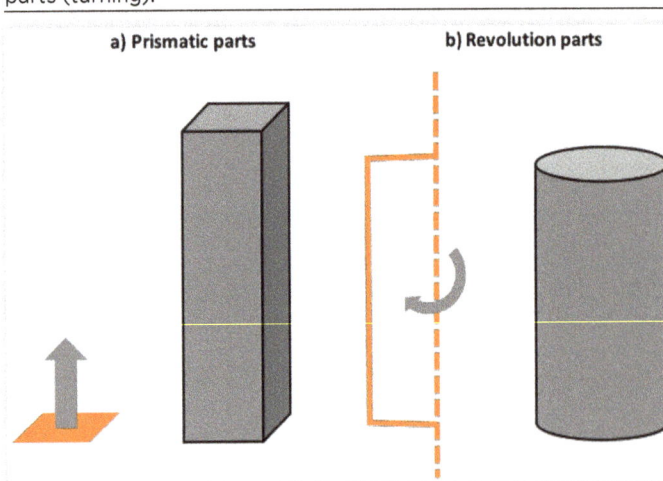

a) Prismatic parts   b) Revolution parts

© Harik/Wuest

we reduce the number of comparably expensive resources (mills) by utilizing lower cost machine tools (lathes) and achieve the same product at a lower price point. While cost is always an important factor, there are other considerations that lead to combining different subtractive processes in a process plan, including qualification of operators, achievable tolerances, and speed.

FIGURE 3.2 Drilling a hole on a lathe.

© Dmitry Kalinovsky/Shutterstock

3.2 Process Planning for Subtractive Manufacturing

Process Planning is the most neglected, yet the most fundamental, process in the manufacturing procedure. The process planner has a reconciliatory role: matching the design with the (available/accessible) manufacturing resources. It requires human analysis of the mechanical part, which is often complex and time consuming. Neglected machining difficulties or unnoticed ones have detrimental consequences on meeting the requirements specified by the part designers. Several research projects (Usiquick, Harik 2007) attempted to automate or to semi-automate Process Planning. However, to this day there is no fully automated process planning tool available that can consistently achieve the same result as a competent human process planner. Nevertheless, the available tools support the process planner in his/her decision and as such are very valuable in themselves. We will see more progress in this area in the next years with the introduction of advanced AI and machine learning techniques that promise to better handle the complexity involved by utilizing the growing amount of manufacturing data captured (see Chapter 10 "Smart Manufacturing").

We list below a few of the core functions a process planner undertakes (Figure 3.3):

1. **Billet Dimensions**
   Building a primary understanding of the part helps the process planner to identify which machines are required for the manufacturing process, e.g., in terms of build volume or maximum diameter. It is a first "rough and simple estimate" of the process and the manufacturing capabilities.

2. **Geometry Analysis**
   Analyzing the geometry of the part helps the process planner to identify which machining modes and tools are required for the manufacturing process. It is a first understanding of the geometry and the eventual manufacturing motion strategies.

3. **Tools Selection**
   A direct result of the geometry analysis function, the process planner connects which manufacturing tools will support obtaining the required geometry with the best possible machining quality. The manufacturing tools selection defines the manufacturing operations and plays a major role in determining the cutting conditions.

4. **Thin or Particular Features**
   A direct result of the geometry analysis function, the process planner localizes particular areas that will require special attention during the machining process. Thin features are an example of where the material thickness is relatively low with respect to its elongation. The machining process, namely the manufacturing roughing stage, will require alternating the material removal from both sides with particular attention to the cutting conditions. Failure to do so will distort the manufactured part and will create flexion due to tool cutting forces.

CHAPTER 3

FIGURE 3.3 CAPP process functions (Harik 2007).

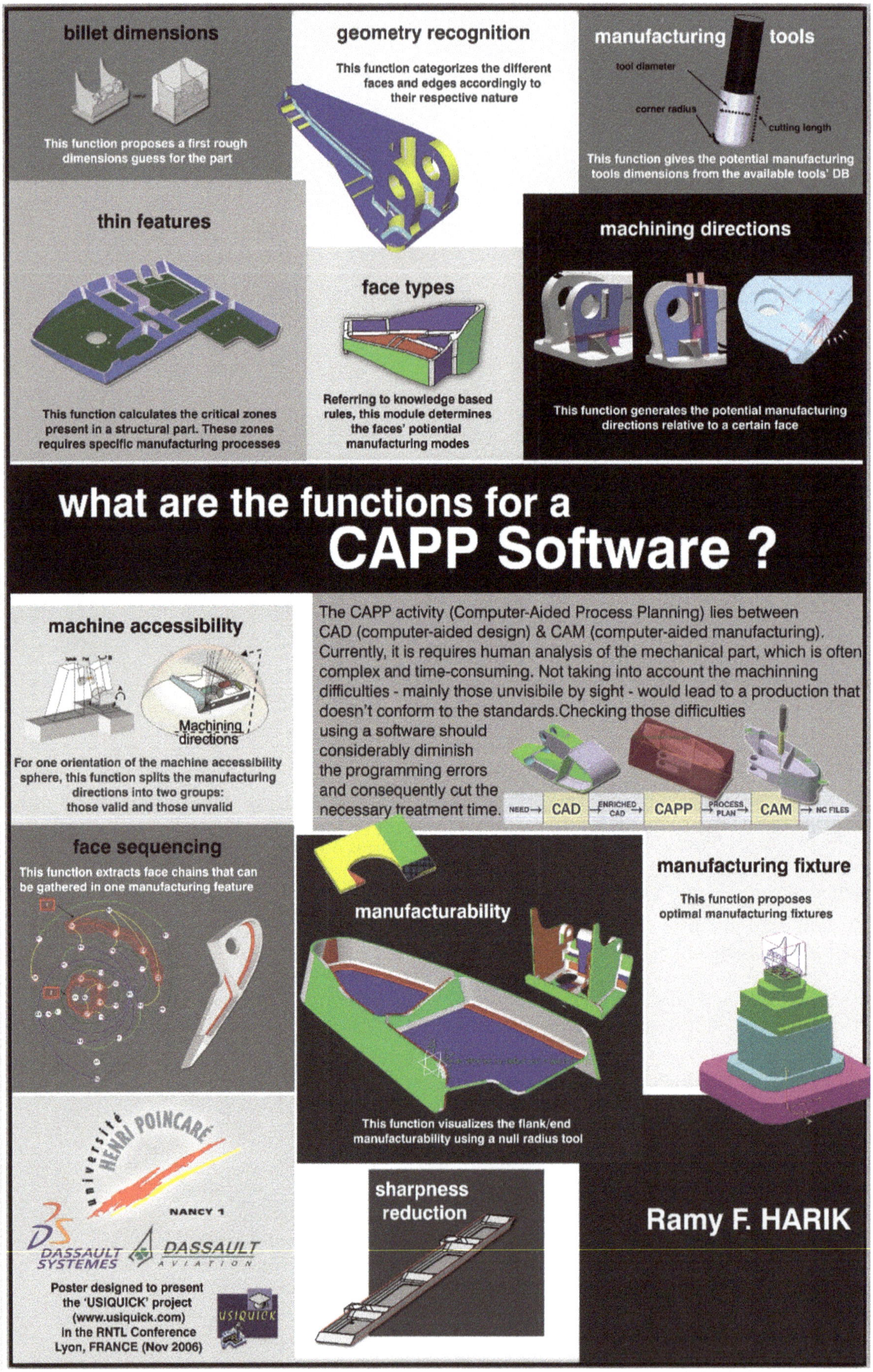

5. **Machining Directions and Accessibility**

   Analyzing the potential machining directions of the shape helps the process planner to identify potential accessibility issues and optimal machining strategies. Machining directions are eventually cross-analyzed to determine the overall ideal strategy.

6. **Machining Sequencing and Optimization**

   A direct result of the machining directions function, the process planner starts grouping successions and sequences in the manufacturing strategy. This helps reduce part marking (surface quality) and the overall manufacturing time as it decreases off-part motions.

7. **Manufacturing Modes**

   The process planner analyses each surface and identifies the manufacturing mode that can manufacture it. This is linked with the selection of manufacturing tools and their parameters.

8. **Fixture Selection**

   A direct result of the machining directions and accessibility, the process planner attempts to identify the manufacturing fixture that will enable the manufacturing process. As a general rule, parts require multiple fixtures and positioning. However, the process planner tries to eliminate as many possible options, without sacrificing the probable quality of the obtained part.

3.3 Milling

Milling is a principal machining form. Figure 3.4 demonstrates the complexity of features that can be achieved with this type of machining processes. Multiple features, such as surfacing and axial features, are common in most mechanical parts. Thin walls and thin bottoms are particular to mechanical parts that undergo weight management or weight reduction, often found in automotive or aerospace applications. Five-axis flank contouring is one of the most complicated features that is manufactured during a milling operation. It cannot be successfully performed with traditional, widely available 3-axis milling machines.

Milling machines are classified according to the degree of freedom they provide to perform the machining operation. The most standard milling machines are 3-axis, 4-axis, and 5-axis ones (Figure 3.5).

3.3.1 Theory of the Orthogonal Model

Eugene Merchant presented the fundamental theory of machining and chip formation in his 1945 publication, listed in the reference section at the end of this chapter. Merchant

FIGURE 3.4 Manufacturing features (Harik et al. 2008).

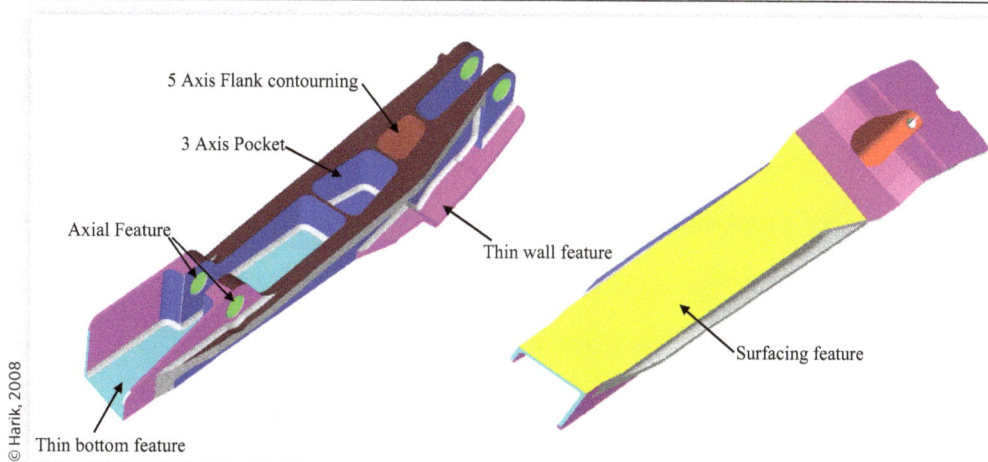

© Harik, 2008

FIGURE 3.5 Milling machine configurations: (left) 3-axis and (right) 5-axis.

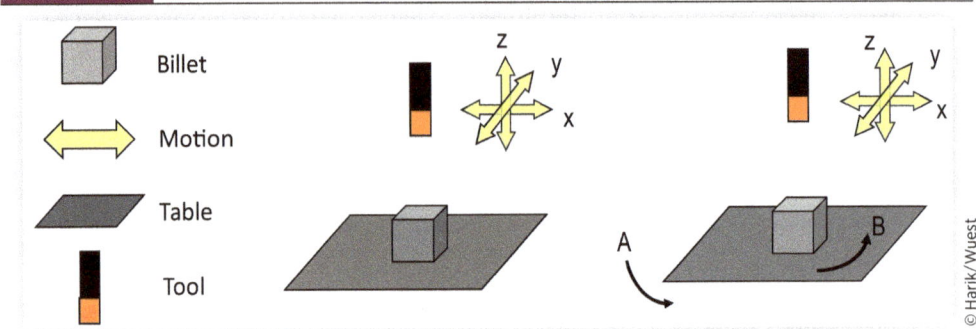

observed the chip formation and, through that process, determined most of the theory that is still valid to date. Recent publications, such as Molinary 2007, demonstrate some discrepancies between Merchant's formulation and obtained experimental data. Such seminal works are important reads for students who desire to deepen their knowledge on machining processes in general. The most current edition of the ASM Handbook, in such scenarios, is a needed resource.

Chip formation theory is based on the orthogonal model. We look at the thrust of one chip off the workpart and study its physics. The chip's nature is an indication of a successful operation with minor defects and markings. The tool penetrates the material at a certain depth of cut defined by the machinist. We typically have two sets of conditions for Roughing and Finishing. Roughing attempts to remove a higher bulk of materials with little to no attention to the surface finishing quality (objective: maximum material removal/time). Finishing targets reduced Rate of Material Removal (R_{MR}) with careful attention to the surface finishing quality (objective: maximum quality).

Figure 3.6 isolates the chip in the context of the material removal process. The chip is in contact with two components: the tool and the part. As the tool is advancing towards the part, the chip is compressed on the tool-chip interface and is thrusted upwards. This movement generates a friction force at the tool-chip interface. The frictional resistance force F is at an angle α from the vertical. The angle is the tool rake angle α. The chip-part interface is defined as the shear plane/shear zone. It is the location where the material finally surrenders to the process physics and allows the shearing of the chip from the part to realize. The angle at which the shearing happens is identified as the shear angle φ. The shear force F_s is located in the shear plane defined by the shear angle φ.

The free-body diagram shows that the chip is subject to four forces:

- F = Frictional resistance (*Friction Force*)
- N = Normal to friction
- F_s = Resistance to shear (*Shear Force*)
- F_n = Normal to shear

FIGURE 3.6 The orthogonal model.

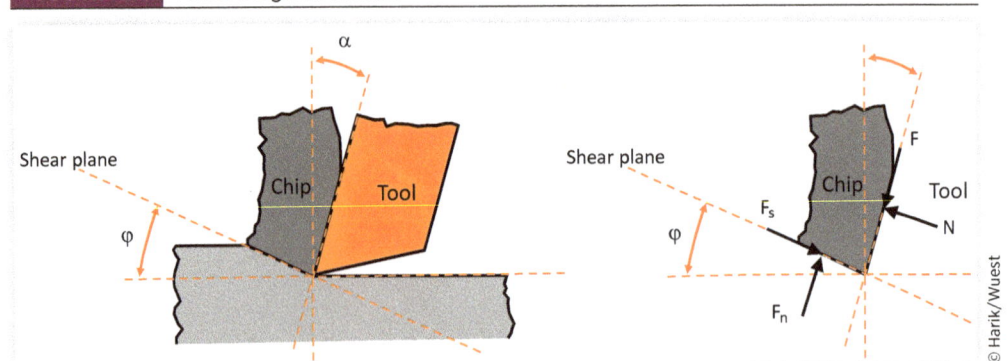

FIGURE 3.7 (Left) Cutting and thrust forces, (right) original chip thickness and final chip thickness.

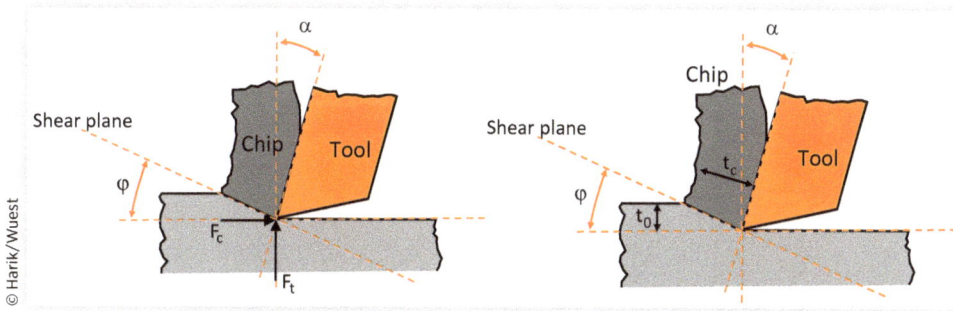

To determine the value of these forces, we mount dynamometers on the tool holders. The dynamometers provide us with the resultant values as well as the horizontal and vertical components. The horizontal component represents actually the Cutting Force F_c. The vertical component represents the Thrust Force F_t that ejects the chip from the part (Figure 3.7).

Using projections along the tool face and the shear plane, we determine the relationship between the different forces.

$$F = F_c \times \sin \alpha + F_t \times \cos \alpha$$

$$N = F_c \times \cos \alpha - F_t \times \sin \alpha$$

Eq. (3.1)

$$F_s = F_c \times \cos \varphi - F_t \times \sin \varphi$$

$$F_n = F_c \times \sin \varphi + F_t \times \cos \varphi$$

Investigating the geometrical setup at the shear plane, we can identify the relationship that connects the shear angle φ, the shear length $\mathbf{L_S}$, and the original chip thickness $\mathbf{t_0}$. The original chip thickness t_0 is equivalent to the depth of cut $\boldsymbol{d}$ ($\boldsymbol{t_0 = d}$).

$$L_s = \frac{t_0}{\sin \varphi}$$

Eq. (3.2)

The computation of the shear length function of the original chip thickness $\mathbf{t_0}$ and the shear angle φ helps to accurately calculate the shear area eventually. This is particularly useful to determine the Shear Stress $\boldsymbol{S}$ endured by the material.

FIGURE 3.8 Computing the shear length.

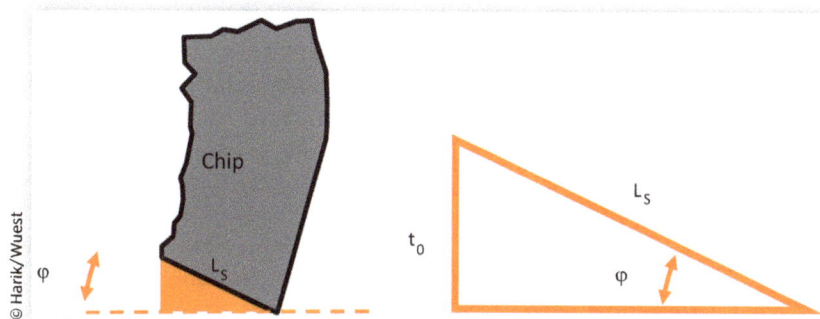

FIGURE 3.9 Computation of the shear area A_s.

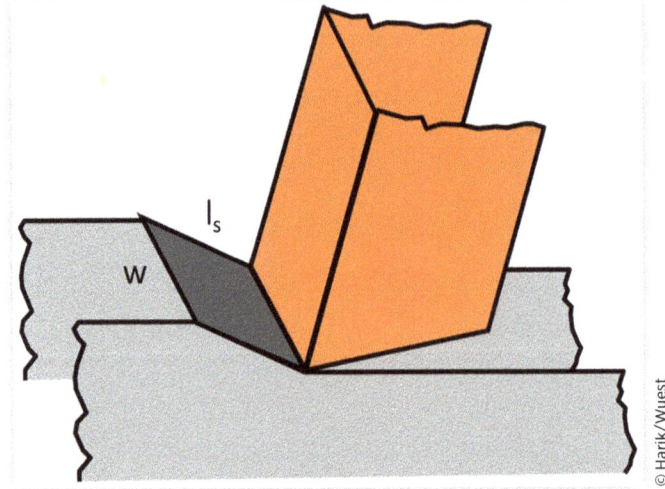

Given that the width of cut is **w**, the shear area $\mathbf{A_s}$ can then be determined as (Figure 3.9)

$$A_s = \frac{t_0 \times w}{\sin\varphi}$$

Eq. (3.3)

And the shear stress **S** can then be calculated as:

$$S = \frac{F_S}{A_S} = \frac{F_c \cos\varphi - F_t \sin\varphi}{\dfrac{t_0 w}{\sin\varphi}}$$

Eq. (3.4)

The coefficient of friction μ at the tool-chip interface and the resulting friction angle λ can be computed with

$$\mu = \frac{F}{N} = \frac{F_c \sin\alpha + F_t \cos\alpha}{F_c \cos\alpha - F_t \sin\alpha} = \tan\lambda$$

Eq. (3.5)

The friction angle is fundamental for the understanding of the *Energy minimization* concept proposed by Merchant. Using a derivation from the shear equations, Merchant proposes the following relationship between the shear angle φ, the tool rake angle α, and the friction angle λ:

$$\varphi = 45 + \frac{\alpha}{2} - \frac{\lambda}{2}$$

Eq. (3.6)

Multiple experiments confirmed that the proposed Merchant equation is indeed a sufficient approximation for some applications. However, Merchant's approximation does not provide an exact computation of the shear angle φ. Therefore, in cases we have the data available to calculate the exact value for the shear angle φ, we generally prefer to not use Merchant's approximation.

Generally, we can distinguish three particular forms of chips as a result of machining operations: continuous chips, discontinuous chips, and built-up edge. *Continuous chips* are often sharp, hot, and, due to their longer form, impose a safety issue to human operators and machine alike. Therefore, we try to avoid the formation of continuous chips as much as possible, e.g., through adding so-called chip breakers on the cutting tool that prevent the chips from becoming long enough to be harmful. Materials that tend to produce continuous chips without repercussions are steel and aluminum, for instance. *Discontinuous chips* are preferred as they imply less safety hazards for machine and machinist, as well as generally produce a better surface finish. Materials producing discontinuous chips are generally harder and more brittle, such as leaded steels and cast iron. The third variation of chips formed during machining is the *built-up edge*, which is technically not ejected as are the other two variants. However, it is material buildup that cumulates at the edge of the cutting tool, and when enough material is collected, it dislocates and often damages the part surface in the process.

3.3.2 Understanding Cutting Conditions

The selection of cutting conditions for any machining operation is crucial and governed by the material of the workpart, the material of the tool used, and the machine power availability. Understanding cutting conditions and the different parameters is crucial for a proper selection and a functional machining process. The principal parameters are tool material, part material, cutting velocity, feed rate, and depth of cut. The *tool material* defines the ability of the tool to withstand the manufacturing operation that is carried. The *part material* dictates the shear strength, and thus the shear stress at which the machining operation is performed. The reader is advised to download an application (such as FSWizard) to access a database connecting work materials and tool materials, and proposed values for the speed and feed.

(Billet) Materials have an optimal value for surface feet per minute *SFM* based on the tool material chosen. Alternative terminology widely used in industry describes SFM also as cutting speed/*cutting velocity V*. The concept is similar to the ability to drive based on your car's performance. If you have a golf cart, you can only go at a certain speed, in contrast to having a sports car depending on the streets surface (racetrack vs. dirt road). The most common tool types are high-speed steel (HSS) tools and carbide tools. Table 3.2 lists standard SFM values in feet per minute for both carbide tools and HSS tools in relation to common part materials.

Once the SFM value is determined based on the machining conditions (i.e., machining aluminum with carbide tools will propose using a SFM value of 600 feet per minute), we can compute the spindle speed **N**. The equation that governs the computation of the spindle speed is provided below:

$$N = SFM \times \left(\frac{12}{Pi}\right) \times \frac{1}{\text{Tool diameter}} \qquad \text{Eq. (3.7)}$$

It is important to highlight that the number 12 in the formula stems from the required conversion of SFM from feet to inches, as the tool diameter is mostly depicted in inches. It is generally advisable to carefully assess the units of the different components used as not only feet/inches but also combinations with metric values pose a challenge and potential problem to the negligent user. The resulting unit for the spindle speed *N* is RPM. While this is a scientific book and we promote the usage of the above formula, it is a common industrial practice to approximate the spindle speed **N** equation by

$$N = SFM \times (4) \times \frac{1}{\text{Tool diameter}} \qquad \text{Eq. (3.8)}$$

TABLE 3.2 SFM – Standard SFM values in feet per minute

Part material	Carbide tool	HSS tool
Aluminum	600	300
Copper	200	100
Cast iron	100	50
Tool steel	100	50
Titanium	100	50
Wood	800	400

© Harik/Wuest

TABLE 3.3 PT – Estimates for 0.05-0.25 in. depth of cut; Data from multiple sources (estimates)

Part material	1/ 8 in.	3/4 in.
Aluminum	0.001	0.006
Copper	0.004	0.008
Cast iron	0.002	0.006
Tool steel	0.001	0.004
Titanium	0.002	0.004
Wood	0.003	0.025

© Harik/Wuest

Machinists in industry typically use the equation above to rapidly compute a potential speed value, or to rapidly confirm a provided value. Therefore, it is important for you to be aware of this in your later careers to not be taken aback by this small deviation in this context.

Following the determination of the spindle speed value **N**, which drives the tool rotation in milling (and the part rotation in turning), we calculate the feed rate f on recommended material-dependent values using the following equation:

$$f = IPT \times n_t \times N \qquad \text{Eq. (3.9)}$$

IPT stands for inch per tooth and is provided by experimental data. Multiple estimates from various sources for material/tool dimension tests are provided in Table 3.3. The IPT is then multiplied by the number of teeth n_t composing the tool. It has to be noted that for both turning and drilling n_t is assumed to be one and thus negligible. In case of turning, utilizing a single-point cutting tool is fairly logical, while for drilling this is a result from previous experimentation. The value of the feed is determined in inches per minute. Chapter 6 "CAD/CAM" focusing on Computer-Aided Manufacturing (CAM) will present the ability to define the feed in the imperial or metric systems.

One important aspect that requires attention of the machining operator and consideration of the production planner is the wear and productive life of the cutting tools under different operating conditions. All cutting tools undergo severe physical stress during operation. While this is expected, the rate of wear and the related potential risk of (unexpected) failure need to be considered. (Regular) Gradual wear of the cutting tool can have an influence on the machining outcome (e.g., surface quality). Failure of a tool represents a more problematic incident for the manufacturer as it may include severe damage to the part, machine, and even wellbeing/safety of the human operator. Therefore the goal is to (a) avoid (unexpected) tool failure and (b) anticipate the rate of wear to allow for timely replacement before the wear affects the quality outcome in a way that is not acceptable. Like in most cases, there is a trade-off to be considered between using the tool as long as possible to save on cost and the risk of producing scrap or parts that require expensive rework.

One accepted way of deriving the productive life of a machining tool was developed by and named after Frederick Taylor – the Taylor's Tool Life equation. The basic form of this equation is given by the cutting speed **V**, the experimentally derived Taylor's tool life constant **C** and Taylor tool life exponent **n** as well as the Tool like **T** and is depicted below:

$$VT^n = C \qquad \text{Eq. (3.10)}$$

Taylor's tool life constant and exponent are both parameters that depend strongly on feed **f**, depth of cut **d**, part material, and cutting tool material. It has to be mentioned, that Taylor's tool life equation has not been theoretically derived but is based on experiments.

In Chapter 10 "Smart Manufacturing," a currently highly researched approach to deal with tool wear and unexpected machine tool failure is presented as part of the predictive maintenance applications.

3.3.3 Rate of Material Removal and Power Requirements

Once the appropriate cutting conditions and tools are determined, we can compute the time and power required to perform the manufacturing operation. These two parameters play an important role in logistical and economic considerations around the manufacturing of a new product/part. Both time and power require the determination of the Rate of Material Removal $\mathbf{R_{MR}}$ (alternative term: Material Removal Rate, *MRR*). Typically, the R_{MR} can be determined as the "uncut area" multiplied by the tool motion towards the material f. The uncut area in milling is determined by multiplying the width of cut w and the original chip thickness $\mathbf{t_0}$ (= depth of cut $\mathbf{d}$) as depicted in Figure 3.10. For milling, we calculate the R_{MR} as

$$R_{MR} = f \times \left(w \times t_0 \right) \qquad \text{Eq. (3.11)}$$

The approximated time required $\mathbf{T_m}$ for the associated manufacturing operation, where $\mathbf{V}$ is the volume of the material to be removed, can then be determined as

$$T_m = \frac{V}{R_{MR}} \qquad \text{Eq. (3.12)}$$

A variation of this method of determining the time required $\mathbf{T_m}$ does not consider the uncut area ($w \times t_0$) and focuses on the length of cut $\mathbf{L}$, mainly due to the fact that the cutting speed in milling is generally constant whether material is removed or not. For a more precise determination of T_m, this method includes a measure $\mathbf{L_A}$, reflecting the "unproductive" movement of the tool (including approach, pre-travel, and over-travel). We present the computation of $\mathbf{L_A}$ when we investigate tool motion in Chapter 6. The resultant formula is depicted as

$$T_m = \frac{L + L_A}{f} \qquad \text{Eq. (3.13)}$$

The power required to perform the operation can be performed through two methods:

- Multiplying the cutting force F_c and the cutting speed V
- Multiplying the R_{MR} by the specific energy of the material u

$$P_c = F_c \times V = u \times R_{MR} \qquad \text{Eq. (3.14)}$$

FIGURE 3.10 Uncut area in milling.

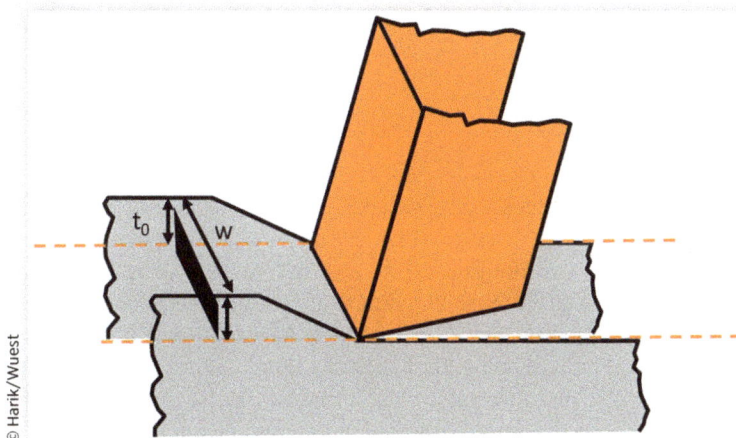

© Harik/Wuest

FIGURE 3.11 Categorization of milling processes.

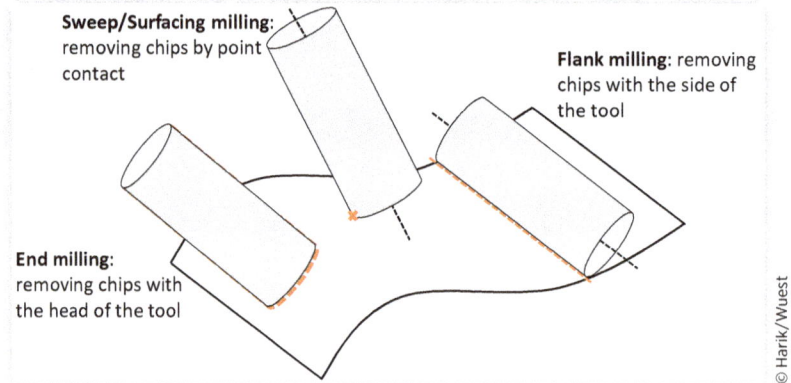

Sweep/Surfacing milling: removing chips by point contact

Flank milling: removing chips with the side of the tool

End milling: removing chips with the head of the tool

The material-specific energy u values are readily accessible for most engineering materials and can be used from commercial sources, such as value tables. Sometimes a correction is needed for the experimentally generated values in the table. Therefore, it is advisable to always use the most current edition of a source such as the ASM Machining Handbook available.

3.3.4 Categorization

Milling is typically categorized according to the combination of tool geometry shape and operation orientation. The same tool can generate different operations, depending on how it is oriented in space. Traditionally, milling is categorized into slitting, plain, helical, slotting, side milling, half-side milling, single angle, double angle, concave, convex, corner rounding, shell end, end milling, key seat, t-slot, and male or female threads according to the ASM Machining Handbook.

While the above categorization is accurate, we prefer the simple macro classification in Figure 3.11 splitting milling operations into end milling, flank milling, and sweep/surfacing milling. *End milling* is when the tool is oriented along the normal of the surface and the chips are removed with the "end" of the tool. *Flank milling* is when the tool is oriented along the isoparametrics (or curvatures) of the surface. Flank milling is famous for the high surface quality and the speed at which it performs the operation. Obviously, tool deflection and accessibility issues hinder the potential of flank milling. Finally, *sweep/surfacing milling* is for all other cutter locations that are not along the normal or the flank of the tool.

It is a common standard today to classify milling processes according to the manufacturing features they represent. Figure 3.12 shows some of these features such as surfacing and pocketing.

FIGURE 3.12 Manufacturing features.

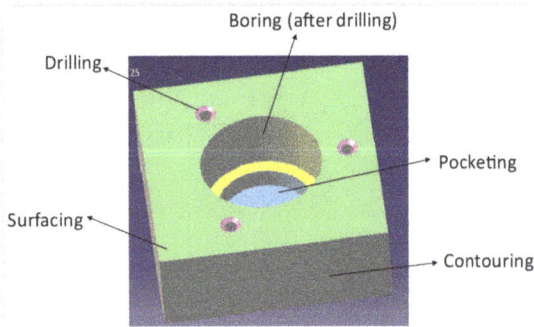

Boring (after drilling)
Drilling
Pocketing
Surfacing
Contouring

3.4 Drilling

Drilling is considered a special milling operation where the tool only moves along its axis, aka a 1-axis milling process (Figure 3.13). The tool plunges into the part and, in its most basic form, generates an axial feature with dimensions equal to the tool diameter. Other axial features are also associated with drilling operations, such as reaming and tapping. They often constitute a subsequent manufacturing operation that follows the initial one. Drilling is known for its reduced complexity, comparably low cost, fast processing speed, and ability

FIGURE 3.13 Drilling motion along the axis of the tool.

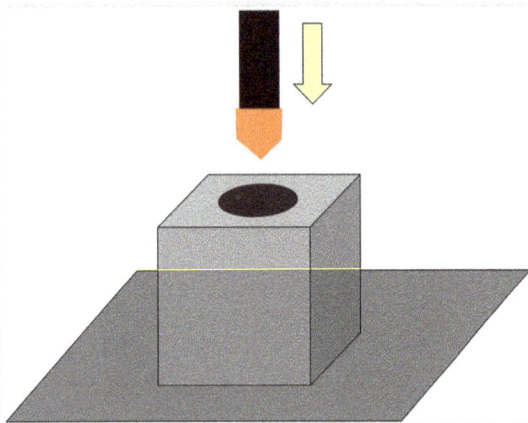

to produce high quality in terms of dimensional accuracy. Furthermore, drilling machinery is readily available in almost every shop floor.

The concept of having multiple manufacturing operations, governed by the type/geometry of the tool used to perform the operation, is in line with the manufacturing program used to document the manufacturing of a part. The manufacturing program is composed of manufacturing phases, manufacturing sub-phases, manufacturing sequences, and manufacturing operations:

- Manufacturing phases are attributed to the selection/identification of a manufacturing machine used as the foundation to conduct an intermediate shape transformation. An example of a manufacturing phase can be the forging station, followed by another manufacturing phase on a milling machine for surface finishing quality.

- Manufacturing sub-phases are identified within a specific manufacturing phase and are attributed to the orientation/fixture selection. Example of a manufacturing sub-phase can be the first surfacing operation on one side of a billet, prior to disassembly and remounting it on another fixture for operations on the other side of the billet.

- Manufacturing sequences are an uninterrupted sequence of operations performed using the same tooling to obtain different manufacturing features.

- Manufacturing operations are attributed to the selection of a manufacturing tool to perform the operation (Figure 3.14). An example of a manufacturing operation is the selection of an end mill to perform a pocketing operation, or the selection of a facing tool for a surfacing operation.

Drilling operations use the same scientific formulations as milling operations with respect to the estimation of forces and power. The computation of speed and feed is also similar; the only difference is that the number of teeth to be used is typically set to 1.

The calculation of the R_{MR} in drilling follows the same principle of feed along the cutting area. Drilling created a cutting area A equal to the area of the cutting tool. The uncut area is a simple circular area dependent on the tool diameter:

$$A = \frac{\pi D^2}{4}$$

Eq. (3.15)

The R_{MR} is then

$$R_{MR} = f \times \frac{\pi D^2}{4}$$

Eq. (3.16)

$$f = IPT \times N$$

Time required T_m for a drilling operation can be computed as the ratio of the removed volume and the R_{MR} as depicted before.

FIGURE 3.14   Manufacturing program schematic (Harik et al. 2013).

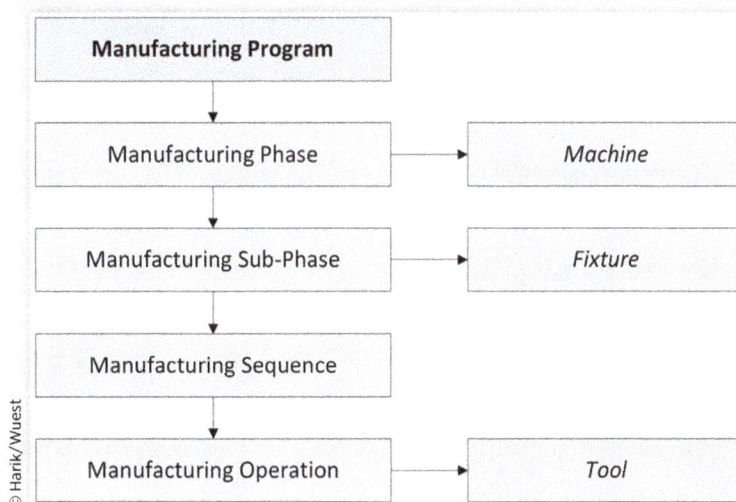

© Harik/Wuest

FIGURE 3.15 Standard drill press commonly available at most shop floors.

Drilling, although being a particular milling type, has been separated for mainly economic reasons. A drill press is comparably cheap to acquire and operate, and prevents blocking a milling machine, requiring a substantial initial investment and higher operating/maintenance cost, to perform the limited drilling operation (Figure 3.15). A very common configuration is the gang-drilling machine where multiple drilling heads are attached to perform the drilling operations on the workpart.

Multiple subsequent operations are possible, following an initial drilling operation. The most common follow-up processes are

- **Reaming**, where we target to slightly enlarge the drilled hole and/or achieve tight tolerances and good surface finish.

- **Tapping** to create internal threads in the manufactured hole.

3.5 **Turning**

The third and final machining process we will focus on in this chapter is turning. Lathe machining performed on a turning center/lathe produces revolution parts. The process is fundamentally different from milling and drilling since the cutting velocity (which dictates the cutting speed) is located at the part itself and not at the tool. In this case, the part turns and is attacked by a stationary cutting tool. While in machining, the tool remains stationary wrt. the cutting speed (the part might however move slightly in a 4- or higher axis setup to accommodate the tool access) and the tool rotates when attacking the part. Moreover, tools in turning operation are single-point cutting tools.

Figure 3.16 demonstrates a typical turning operation. The circular motion is conducted on the part and the movement of the turret; the tool holder provides the linear motion. Turning operations are also simpler to define as the tool motion is described in a plane rather than in three dimensions. Lathe machines are often referred to as 2-axis CNC or 2.5-axis CNC. The justification used by entities labeling lathe as 2.5 is that the motion of the part, which creates the 3D shape, is considered as half an axis since it cannot be controlled to produce shapes.

Configuration of lathe machine is typically separated into the X and Z axes. The Z axis follows the Z convention which will be explained in Chapter 6 on computer-aided manufacturing. Basically, there exists a reference location where Z can be set to 0, separating the manufacturing domain into two options: Material exists potentially, Material does not exist. The Z convention helps operators to anticipate machine engagement and to properly be aware of the manufacturing situation (Figure 3.17).

FIGURE 3.16 Turning operation.

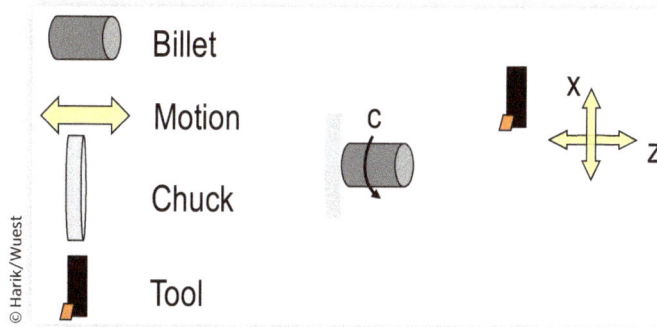

FIGURE 3.17 Lathe standard axis.

There is a conversion from the orthogonal model presented under milling to turning. The fundamental concept remains the same however we try to match the attributes. This means the original depth of cut in milling t_0 (= d) stands for the feed rate f in turning, while the width of cut w in milling stands for depth of cut d in turning. Groover presents a detailed graph for the approximation of turning by the orthogonal model that the reader can refer to for further information (Groover 2010).

The R_{MR} in turning follows the same fundamental concept of the uncut area multiplied by the tool movement. It is important to highlight that the revolution of the part automatically doubles the depth of cut on the final part. The relationship between the starting diameter D_o, the final diameter D_F, and the depth of cut d is provided by

$$d = \frac{D_o - D_F}{2}$$
Eq. (3.17)

The equation above can be generalized to include the number of passes n, and is useful to compute the final diameter following a roughing/finish operation (Figure 3.18).

$$D_F = D_o - 2 \times n \times d$$
Eq. (3.18)

The uncut area is presented as the dark gray area on the right-hand side in the picture above. It depicts the difference between the original circular cross section of diameter D_O and the final circular cross section of diameter D_F. The uncut area can be calculated by

$$A = \frac{\pi D_O^2}{4} - \frac{\pi D_F^2}{4} = \frac{\pi}{4} * \left(D_O^2 - D_F^2 \right)$$
Eq. (3.19)

FIGURE 3.18 Depth of cut in turning and its effect on the final part.

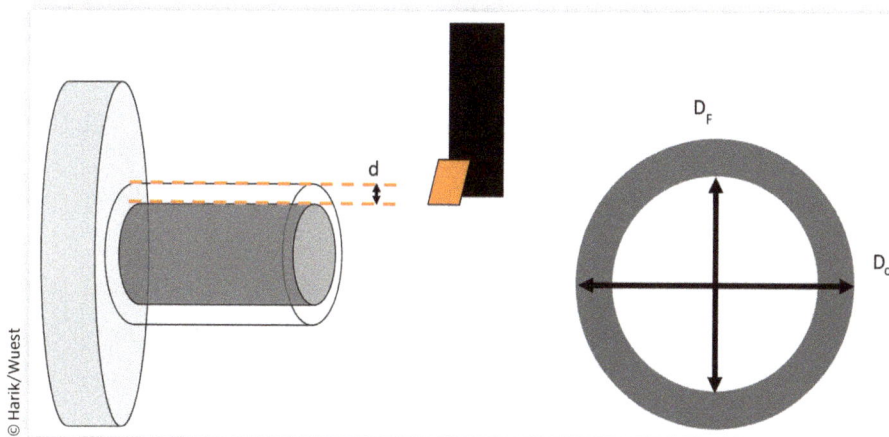

The R_{MR} is given by

$$R_{MR} = f * \left(\frac{\pi D_O^2}{4} - \frac{\pi D_F^2}{4} \right)$$ Eq. (3.20)

While turning operations often focus on the removal of material from the outer dimensions of revolution parts (e.g., manufacturing a baseball bat), turning allows for additional processing that targets, e.g., the internal diameter of a ring. A common process to enlarge an existing hole in a revolution part on a lathe is boring. The boring process is based on a single-point cutting tool similar to regular turning operations, and can be used to achieve a better surface finish, tighten tolerances, center the hole along a rotational axis, or put a slot in the part (grooving).

References

Groover, M., *Fundamentals of Modern Manufacturing: Materials, Processes, and Systems*, 4th ed. (Hoboken, NJ: John Wiley & Sons, Inc., 2010).

Harik, R., "Spécifications de fonctions pour un système d'aide à la génération automatique de gamme d'usinage: Application aux pièces aéronautiques de structure, prototype logiciel dans le cadre du projet RNTL USIQUICK, Doctoral dissertation, Université Henri Poincaré-Nancy I, 2007, https://tel.archives-ouvertes.fr/tel-00173161/.

Harik, R.F., Derigent, W.J., and Ris, G., "Computer Aided Process Planning in Aircraft Manufacturing," *Computer-Aided Design and Applications* 5, no. 6 (2008): 953–962.

Harik, R. F., Gong, H., and Bernard, A. (2013). 5-axis flank milling: A state-of-the-art review. *Computer-Aided Design*, 45(3), 796–808.

Merchant, M.E., "Mechanics of the Metal Cutting Process, I: Orthogonal Cutting," *Journal of Applied Physics* 16 (1945a): 267–275.

Merchant, M.E., "Mechanics of the Metal Cutting Process, II: Plasticity Conditions in Orthogonal Cutting," *Journal of Applied Physics* 16 (1945b): 318–324.

Additive Manufacturing

Additive Manufacturing (AM) represents processes where the material is transformed from Shape A to Shape B by the addition of material. The fundamental concept of AM is that we augment the volume of materials throughout the process. The starting form is mostly "nothing," aka zero starting material/volume. However, while still under development and not very common, there are specific processes that allow us to add material to already existing structures, such as direct energy deposition processes and others, such as ultrasonic consolidation. The latter can technically be considered a hybrid process, since we add and remove material. This chapter will introduce selected innovative and widely used AM processes such as Stereolithography (SL), Selective Laser Sintering (SLS) (see Figure 4.1), and Fused Filament Fabrication (FFF). AM is often classified based on the process and the type of starting material form/state, and not based on the geometrical form we desire to produce.

AM uses materials, machinery, and tools to build up material layers or material zones gradually, which ultimately cumulates in the creation of the final part. Traditionally, this buildup process was performed with parallel layers, most commonly referred to as slices. Recent advances in AM enable the out-of-plane creation of parts. These parts are self-supported and directly optimized with respect to their directional properties. Similar to all other manufacturing processes, AM is an economical process where we augment the monetary value of the raw material through the transformation stage.

The AM process generally undergoes a number of pre-processing and post-processing steps to achieve the desired outcome. Several steps are taken to ensure the manufacturing process is smooth and with minimal interruptions and defects. We often categorize the generalized AM process into (1) Engineering Design, (2) Pre-Processing, (3) Manufacturing, and (4) Post-Processing. In the first step, the *Engineering Design* must consider the specific limitations of the equipment and resources available. Functions such as the support structure and minimum features, which are discussed in the subsequent paragraphs, must be accounted for. During the following *Pre-Processing* step and depending on the chosen AM format, we extract the layered

FIGURE 4.1   Sample of AM process from the raw materials to the final part.

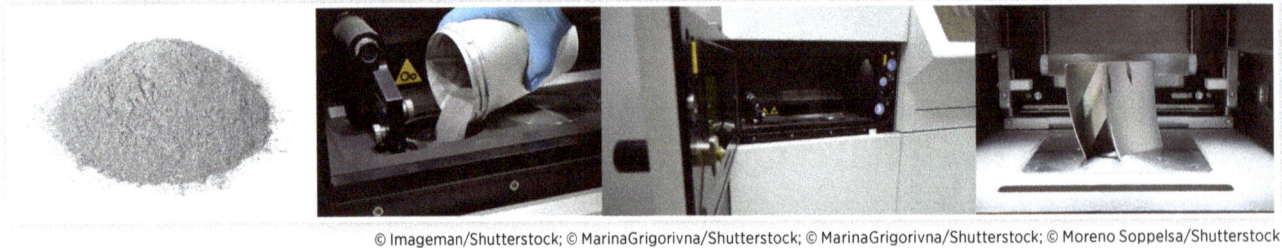

© Imageman/Shutterstock; © MarinaGrigorivna/Shutterstock; © MarinaGrigorivna/Shutterstock; © Moreno Soppelsa/Shutterstock

FIGURE 4.2   Generalized process for AM.

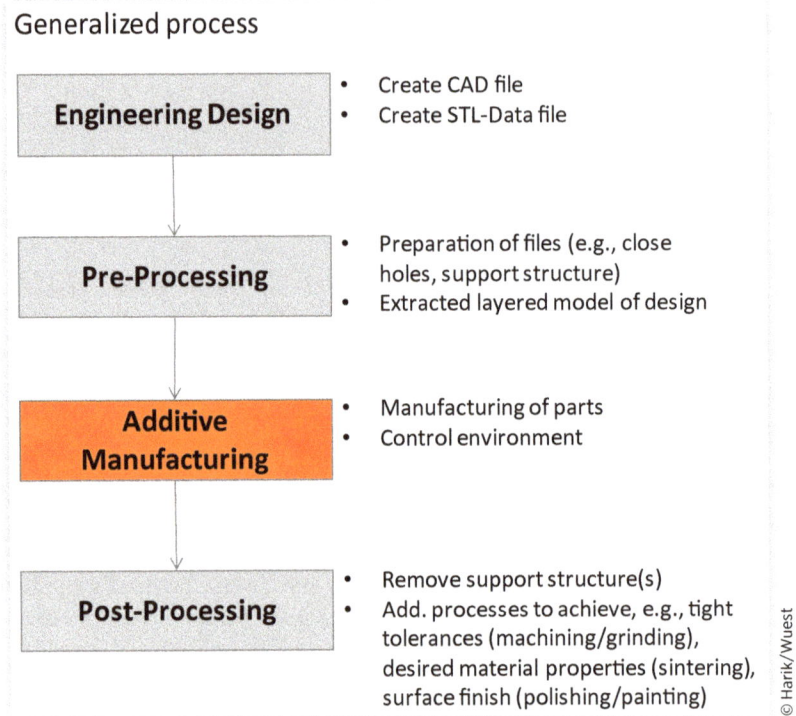

Generalized process

Engineering Design
- Create CAD file
- Create STL-Data file

Pre-Processing
- Preparation of files (e.g., close holes, support structure)
- Extracted layered model of design

Additive Manufacturing
- Manufacturing of parts
- Control environment

Post-Processing
- Remove support structure(s)
- Add. processes to achieve, e.g., tight tolerances (machining/grinding), desired material properties (sintering), surface finish (polishing/painting)

© Harik/Wuest

model and generate the "slices." Slicing is the most basic form used to perform the local fusion process in AM. The heat and/or consolidation source will then follow the proposed slices to build the part layer by layer. The third step, the (Additive) *Manufacturing* of the design, depends largely on which process and technology was selected. In this book, we detail three distinct and widely used processes within this chapter, and elaborate on a fourth, Advanced Fiber Placement, in the later Chapter 8 "Composites Manufacturing." The last step, *Post-Processing*, focuses on removing any support structure that might have been needed during the manufacturing phase and also may include manipulation of the newly created part through material removal processes such as grinding/cutting. This might be necessary to ensure specific requirements towards surface quality or tolerancing that is not (yet) achievable with AM processes. Figure 4.2 presents the generalized process for AM.

Currently, AM is considered a highly flexible manufacturing process which is suitable for one-of-a-kind and small-scale production runs (Figure 4.3). It presents no signification upfront cost, other than the system itself. This is an important differentiation when we think of traditional manufacturing processes that requires extensive and elaborate tooling and/or molds. This is however not an indication that AM processes are not expensive: materials used for the AM process are still expensive and, additionally, additively manufactured parts do not hold a competitive advantage when it comes to several important part/product properties, e.g., strength (with the exception of some metal processes). To manufacture truly functional parts with given strength requirements, Subtractive Manufacturing is still the preferred method today. Recent advances in composites three-dimensional (3D) printing and Composites Manufacturing in

FIGURE 4.3 AM is capable of producing small and large scale parts/products.

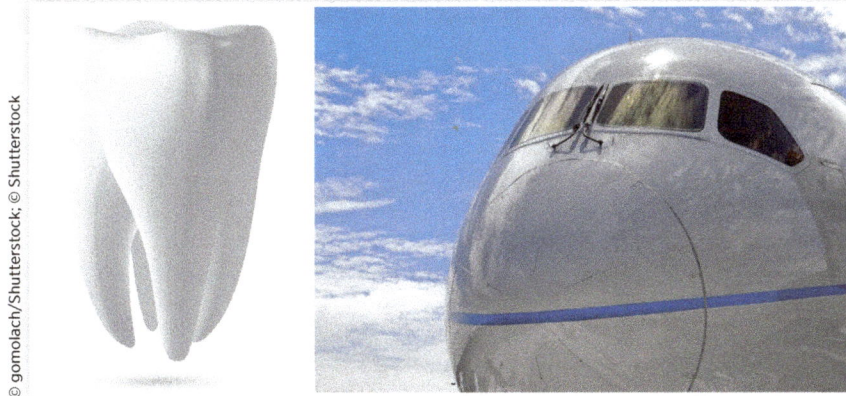

© gomolach/Shutterstock; © Shutterstock

general is however making steady steps towards overcoming this preference, as well as metal additive processes and bioprinting. One prime example of the potential that AM offers is the a-CT7* engine mid-frame redesign by GE Aviation (Figure 4.4). The new design of the engine mid-frame is manufactured using AM and enabled a reduction of parts from ~300 to 1 (7 assemblies to 1) and an improved performance. Furthermore, by reducing the complexity and number of parts, the redesign had a tremendous impact on the whole value chain as it reduced the number and tiers of suppliers, necessary coordination, and stockholding, as well as several other aspects. This example shows the opportunities that AM provides when applied (1) well and (2) for a suitable part/assembly.

Table 4.1 depicts selected general advantages and disadvantages of AM. It has to be noted that this grouping strongly depends on the selected AM process (especially for the items marked with an "*" in the table).

One of the biggest advantages of AM is its high flexibility. This does not come as a surprise, since AM processes were first conceived as ways to Rapidly Prototype, so that designers can quickly verify a certain shape or feature, prior to planning and launching full-scale production. Figure 4.5 outlines the high flexibility of AM in comparison to other commonly used, traditional manufacturing processes from the Deformative and Subtractive Manufacturing families.

AM is a relatively new manufacturing field in comparison with Deformative and Subtractive Manufacturing. The development of computer graphics and computer-aided design (CAD) systems was a key enabler for AM and rapid prototyping progression. The three major AM processes detailed in this chapter – Stereolithography, Fused Deposition Modeling (FDM, currently labeled as Fused Filament Fabrication), and Selective Laser Sintering – were patented in 1984, 1989, and the early 1990s, respectively. Figure 4.6 details the major milestones on a

FIGURE 4.4 The a-CT7 engine mid-frame (Source: GE Aviation).

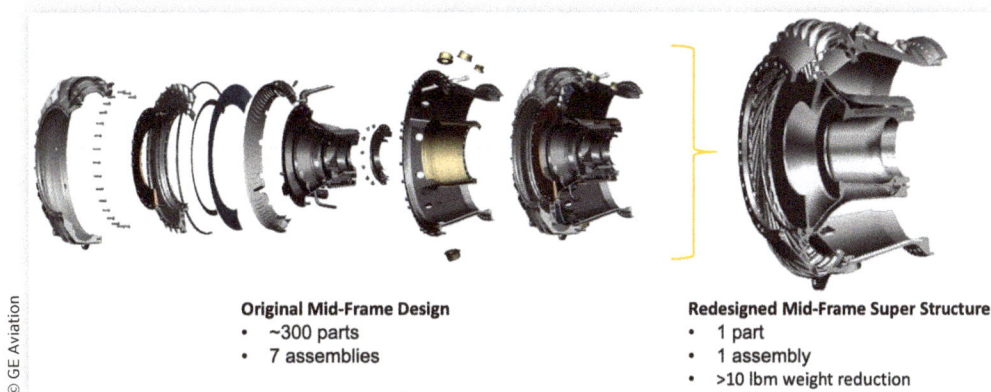

© GE Aviation

Original Mid-Frame Design
- ~300 parts
- 7 assemblies

Redesigned Mid-Frame Super Structure
- 1 part
- 1 assembly
- >10 lbm weight reduction

---

* https://www.geaviation.com/commercial/engines/ct7-engine

CHAPTER 4

TABLE 4.1 Generalized advantages and disadvantages of AM

Advantages	Disadvantages
• High degree of flexibility	• Not economical for large batch sizes
• Ability to manufacture complex, near-net shapes	• High manufacturing time required*
• Ability to manufacture assemblies	• Skill and expertise needed to prepare design
• Low ramp-up investment	• Requires post-processing (e.g., removal of support structure for some)
• Reduced lead time	• Materials available limited and expensive (improving!)
• Ability to manufacture multi-material/multi-color parts*	• Systems (and maintenance) cost is high (improving rapidly!)
• Excellent mechanical properties*	• Limited mechanical properties for some (counterexample: titanium and chromium alloys have excellent properties)*
• Little/no wasted material	• Some waste materials can be hazardous*

© Harik/Wuest

FIGURE 4.5 Comparison of flexibility of manufacturing processes with AM.

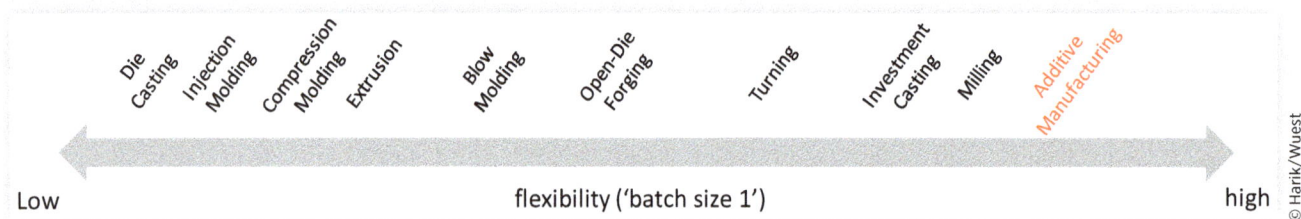

FIGURE 4.6 Evolution of AM.

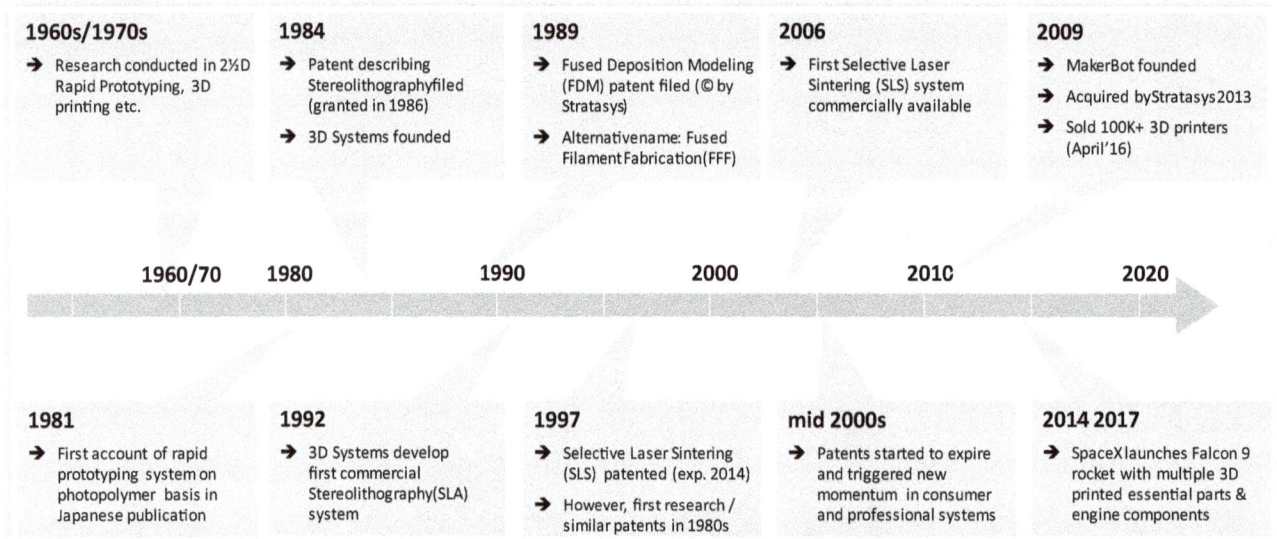

timeline illustrating the evolution of AM, including selected major achievements that can be seen as an indication of the level of maturity of the industry.

This chapter, following this extensive introduction, is outlined as follows. First, we present our AM materials classification system. Unlike in Subtractive Manufacturing, AM is not classified based on the input or desired (output) shape but rather on the starting material and desired properties of the final part. In Sections 4.2, 4.3, and 4.4, we selected three of the seven AM classifications according to the American Society for Testing and Materials (ASTM) Committee F42: Material Extrusion, Powder Bed Fusion, and Vat Photopolymerization. We present each of these three AM processes in detail and highlight their differences and individual advantages. We discard Binder Jetting, Direct Energy Deposition, and Material Jetting as those are either not very common, their application area is rather limited, or they are very similar to one of the chosen three in their theoretical set-up. It is worthwhile to note that ARCAM and LENS are pretty successful in metal AM, and these will be detailed in future editions. Sheet Lamination is extensively covered in the later "Composites Manufacturing" chapter. We conclude our chapter with Sections 4.5 and 4.6 on Process Planning and the challenges in AM.

4.1 **Material Classification**

The most common classification of AM processes can be made with respect to the starting material. It can be safely stated that there is a direct correlation between the AM process and its starting material. Figure 4.7 demonstrates an abstract view of AM and the different material starting points.

The ASTM International Committee F42 on Additive Manufacturing Technologies defines AM as *"A process of joining materials to make objects from 3D model data, usually layer upon layer, as opposed to subtractive manufacturing methodologies."* The Committee defines seven different AM process categories:

- **Binder Jetting**: *"[...] liquid bonding agent is selectively deposited to join powder materials."*

- **Direct Energy Deposition**: *"[...] focused thermal energy is used to fuse materials by melting as they are being deposited."*

- **Material Extrusion**: *"[...] material is selectively dispensed through a nozzle or orifice."*

- **Material Jetting**: *"[...] droplets of build material are selectively deposited."*

- **Powder Bed Fusion**: *"[...] thermal energy selectively fuses regions of a powder bed."*

- **Sheet Lamination**: *"[...] sheets of material are bonded to form an object."*

- **Vat Photopolymerization**: *"[...] liquid photopolymer in a vat is selectively cured by light-activated polymerization."*

While AM provides a new level of "design freedom," there is yet the material challenge to overcome. AM processes are classified based on the starting materials in the process:

- **Filaments** that are introduced by an extrusion head, and fused together constitutes family of **FFF**

- **Powders** that are spread and fused together using a laser or electron beam constitutes family of **SLS**

- **Liquid photopolymers** that are available in a basin and are cured into solid polymers by means of laser constitutes the family of **SL** (sometimes referred to by the machine producing it as Stereolithography Apparatus [SLA])

FIGURE 4.7 Abstract view of AM.

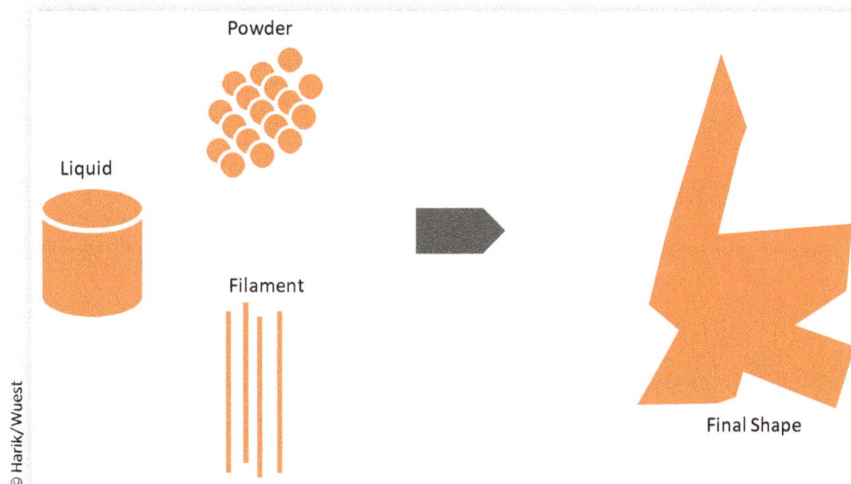

© Harik/Wuest

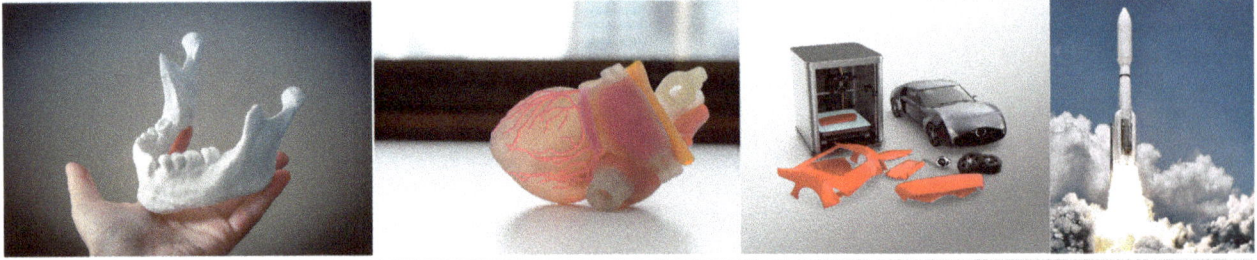

© MarinaGrigorivna/Shutterstock; © everytime/Shutterstock; © Chesky/Shutterstock; © Shutterstock

An important aspect is the application areas, which spans from healthcare to aerospace (Figure 4.8). This multi-domain applicability is what enables AM to gain rapid traction on Subtractive and Deformative Manufacturing.

4.1.1 Filaments

Filaments are widely available in a range of different materials, qualities, and prices. Common filament materials are PLA (Polylactic Acid) and ABS (Acrylonitrile Butadiene Styrene). There exist multiple other filaments such as Nylon, ULTEM, PET (Polyethylene Terephthalate), Polypropylene (PP), and several specialized ones like reinforced nylon, etc. Several research facilities are developing their own materials that respond to their specific needs and requirements such as aerospace grade materials (ULTEM) that are fireproof or biological material that are chemically inert. Filaments come in different sizes and are often based on the extrusion head of the FFF apparatus (Figure 4.9). Most commonly available sizes are filaments with a 1.75 mm or 3 mm diameter. Filaments are considered one of the cheapest and most available forms of AM raw materials. You can actually get a selection of filaments for FFF at general stores or online, which is an indication for just how common FFF is nowadays. Furthermore, this also showcases that the low-cost FFF systems are not only used by industry or other professionals any longer but also increasingly by the "Average Joe" at home, in their workshop or garage.

FIGURE 4.9 Filaments are often associated with FFF (Fused Filament Fabrication).

© MarinaGrigorivna/Shutterstock

4.1.2 Powder

Powders as a raw material for SLS are increasingly available yet still expensive (Figure 4.10). Additionally, in contrast to most filaments, metal powders require some safety-related arrangements and additional systems (e.g., glove box or special vacuum cleaner) when handling them due to their particle size (=< nm possible) and reactive nature of some (e.g., magnesium). Some SLS apparatus are linked to a unique set of materials, creating a kind of monopole on specific processes. Moreover, SLS raw materials need to be purchased in preset quantities, making it harder to experiment with different materials to obtain different properties. Also, typically the SLS process needs to be performed in a confined environment to meet certain safety precautions. Common metal powder variations are aluminum alloys, cobalt-based alloys, nickel-based alloys, stainless steel/tool steel, and titanium alloys. While typical variations are metal based, there are other common variations available such as ceramics (e.g., Silicon Carbide) and polymers (e.g., Ultrasint PA6 - X028).

4.1.3 Liquid Photopolymers

Used for SL, liquid photopolymers are typically a mix of monomers with oligomers and photoinitiators. Oligomers are a few units of monomers that will be joined with the monomers once an ultraviolet (UV) light initiates the process. Similar to SLS, SL raw materials are rather costly; however their prices also have a decreasing tendency. For **liquid photopoly-mers**, a wide variation of specialized materials are trade-marked and available only from the respective provider Ceramal (© 3D Systems) ceramic-reinforced composite, EPU60 (© Carbon) Elastic Polyurethane, and 18420 by SOMOS, which allows the creation of RTV molds (Figure 4.11).

FIGURE 4.10 Powders are associated with SLS (Selective Laser Sintering).

© Imageman/Shutterstock

4.2 Fused Filament Fabrication

Fused Filament Fabrication (FFF) is the scientific adopted name for Fused Deposition Modeling (FDM) as the later is patent protected by Stratasys©. Scott Crump, founder of Stratasys, developed the first prototype in 1989. The idea apparently came to him as he was creating a toy frog for his daughter using a simple glue gun. The development accelerated well into the 1990s, when the first commercial systems became available. Another major milestone was achieved by FDM/FFF when it became the most widely available AM process in 2003. In 2012 and 2013, respectively, Stratasys merged in Objet and acquired MakerBot, underlining their leading position in the FFF market. In the following, we present the basic understanding of the process, followed by the current tool of practice and application venues.

4.2.1 Process

Definition from ASTM International F2792-12a: Fused Filament Fabrication is "a material extrusion process used to make thermoplastic parts through heated extrusion and deposition of materials layer by layer." FFF processes are widely in use due to their relatively low initial cost of the system, availability of low cost materials, and low maintenance they require. The FFF process drives a filament raw material from a spool(s) through an extruder that heats/melts the filament and, using a deposition platform, deposits the material precisely where needed. The part is being built layer by layer on the platform, which is often a heated surface to allow for material tackiness. Traditionally FFF processes were conducted in contained environment,

FIGURE 4.11 Liquid photopolymers are associated with SL (Stereolithography).

© MarinaGrigorivna/Shutterstock

CHAPTER 4

FIGURE 4.12 FFF/FDM process basics.

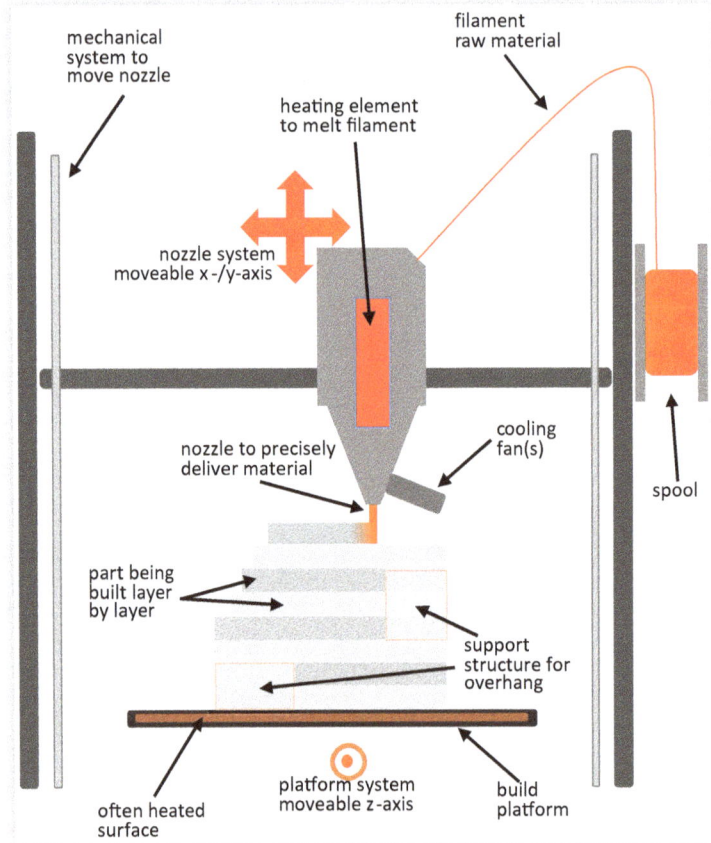

© Harik/Wuest

but recently open-chamber FFF tools are more and more common. Open-chamber FFF systems have the disadvantage that extreme environments, such as extremely cold labs due to A/C, can cause quality issues with the prints. The motion of the extruder/platform is performed with mechanical motors operated by numerical control commands. There are different variations of systems available with different degrees of freedom for platform and extruder head. In Figure 4.12, we depict on a possible setup, where the platform moves on the z-axis, while the extruder operates on the x- and y-axis. In alternative systems, the platform might move on the y-axis while the extruder is moveable on the x- and z-axis. However, the principle function of the system is yet the same.

4.2.2 Tools

Overall FFF is comparably low cost/low maintenance, and thus most widespread (Figure 4.13). The most common and available FFF system is the MakerBot Replicator in all its variations. It

FIGURE 4.13 FFF tools.

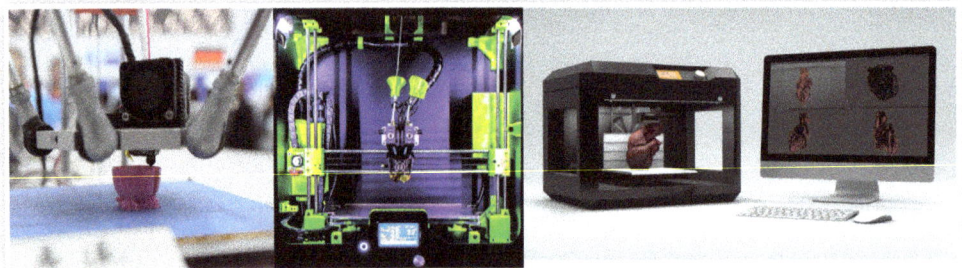

© enmyo/Shutterstock; © Jonathan Lao/Shutterstock; © Denis Ronin/Shutterstock

FIGURE 4.14 FFF parts/products.

FIGURE 4.14 FFF parts/products.

© ShooLandia/Shutterstock; © Photo Oz/Shutterstock; © Glacee/Shutterstock

can be bought "everywhere," e.g., Sam's Club (2016). We often characterize tools (aka 3D printers) with respect to their resolution, building volume, and material variety. Naturally, tool cost and operation cost are a major factor when it comes to selecting a system.

Commercial systems possess better resolution, building volume properties, and a more enhanced variety of engineering materials. Naturally the commercial systems have higher throughput, however they are far more expensive.

4.2.3 Applications

FFF can produce a variety of parts and products with a broad range of applications, ranging from designs to prototypes and functional parts (Figure 4.14).

Industrial	Research
• Automotive: transparent front/back light covers, mirror casings, etc.	• Rapid tools/fixtures for lab experiments
• Aerospace: lightweight composite (complex) parts; production fixtures	• Visualize theoretical concepts; replicate rare specimens (e.g., dinosaur skull)
• Manufacturing: job-shop production; tooling; fixtures	
Education	**Healthcare**
• Visualize 3D designs in STEM programs	• Individualized support
• Support for projects for students (e.g., robotics)	• Low-cost, rapid products (e.g., bionic hand)

4.3 SLS

In 1989, SLS was invented and patented by DTM. The DTM SinterStation was the first SLS system that was commercially successful in 1992 and is often considered the pioneer of SLS machines. In 2001, 3D Systems acquired DTM; however, it has to be noted that the SLS patent expired in 2014. Similar to the expired patent on FFF, this spurred several new systems to enter the market, including a desktop version. However, SLS systems are significantly more complex than FFF systems and also impose stringent safety measures that somewhat exclude these systems to some extent from a wide adoption by the "Average Joe" to this day.

4.3.1 Process

Definition from ASTM F2792-12a: Selective Laser Sintering is a "powder bed fusion process used to produce objects from powdered materials using one or more lasers to selectively fuse or melt the particles at the surface, layer by layer, in an enclosed chamber." Other commonly used and closely related powder bed fusion processes are Selective Laser Melting (SLM) and Direct Metal Laser Sintering. Historically, SLS and SLM are often used to describe the same process;

FIGURE 4.15 SLS process basics.

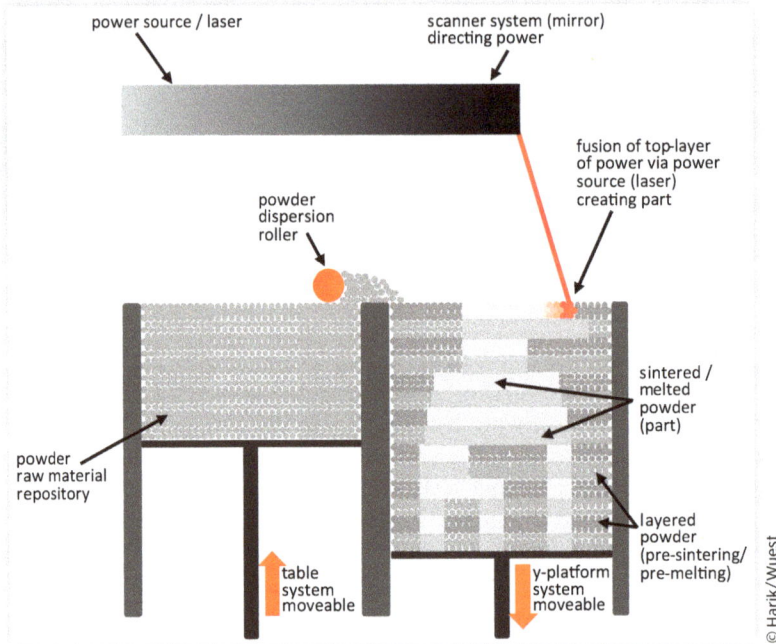

however, some argue that SLS is associated with polymer/ceramic powders while SLM focuses on metal powders, again highlighting the connection of process to raw material in AM process classification.

The SLS process uses a powder dispersion system to create a raw layer of powder (Figure 4.15). Then, a power source, in most cases a laser system, transmits heat to fuse the top layer creating a "slice" of part. Layer by layer, the whole part is built by repeating the process of (1) replenishing a layer of powder and (2) fusing where needed. The raw material is typically on a platform that travels downward as we disperse a new layer of raw power. Figure 4.16 details the general SLS process.

4.3.2 Tools

SLS tools are some of the most expensive and complex variations within today's available AM processes. The original investment cost for commercial systems can go between one quarter and more than a million USD. Moreover, SLS tools require special maintenance that is valued (according to NIST) around $30k/year. Noncommercial tools are not common; however, most likely associated with the original SLS patent recently expiring, ongoing research/breakthrough for lower-cost SLS tools are currently in release. SLS generally have

FIGURE 4.16 SLS tools.

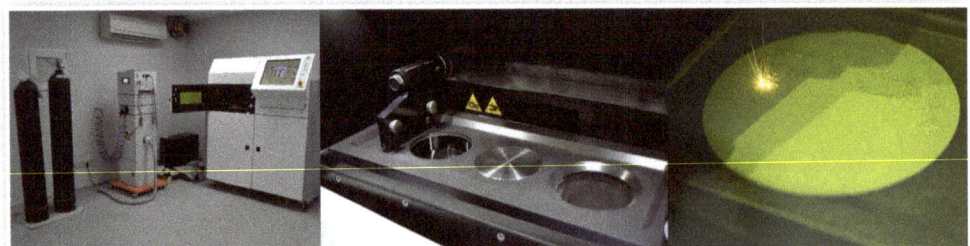

© MarinaGrigorivna/Shutterstock

a medium build area with a limited selection of materials; however they offer a good resolution and layer thickness.

4.3.3 Applications

SLS allows a wide variation of materials, (surface) finishes, and high accuracy; thus it leads to a wide application area (Figure 4.17). It is possible to directly create fully functional parts/products from metal, ceramic, or polymer with good material properties. Table 4.2 lists the most dominant industrial and research applications of SLS.

FIGURE 4.17 SLS part/product.

© Moreno Soppelsa/Shutterstock

4.4 Stereolithography

In 1984, Charles Hull patented the SLA process in the USA. Charles Hull is the founder of 3D Systems who later acquired DTM (and thus SLS) in 2001. The first commercial system, 3D Systems SLA-1, was released in 1987. The SL process has evolved from 1D to 2D to 3D … heat transmission enabling to instantaneously obtain the 3D model. SL is often labeled as SLA which is misleading. SL is the process standing for stereolithography. SLA stands for the tool enabling the SL process. SLA is the Stereolithography Apparatus. In the following sections we detail the process, tools, and applications of the SL process.

4.4.1 Process

Definition from ASTM F2792-12a: Stereolithography is "a vat photopolymerization process used to produce parts from photopolymer materials in a liquid state using one or more lasers to selectively cure to a predetermined thickness and harden the material into shape layer upon layer." SL processes can manufacture products/parts with high surface quality from different and varied materials. Originally, SL was mainly used for design and/or test prototypes. However, there are current applications where fully functional parts/products are produced with SL.

The SL process is similar in nature to other AM processes in the concept that we build/augment the part layer by layer. The layer fabrication process in SL is through the curing of the top-layer liquid photopolymer via a power source, most often a UV Laser. Power channeling modes have evolved from a single point (circle) to a linear form all the way to the current state of the art with research on 3D curing waves capable of creating the 3D shape without the layer-by-layer process. Figure 4.18 demonstrates the general SL process basics.

4.4.2 Tools

SL are more "desktop" available then SLS tools, however, far less than FFF ones (Figure 4.19). While systems were previously expensive with an associated high maintenance cost, new desktop variants (i.e., Formlabs Form 1+/Form 2) are more widely available today. However, the required materials are expensive and their availability is limited. Commercial and industrial systems

TABLE 4.2 SLS industrial and research applications

Industrial	Research
• Automotive: Engine parts; structural parts, etc.	• Test prototypes of new innovative structures, etc.
• Aerospace: Integral propulsion system parts, etc.	• New material/design combinations, etc.
• Manufacturing: Tooling; fixtures; etc.	
• Fashion: Clothes/shoes, e.g., Nike Vapor Laser Talon	

CHAPTER 4

FIGURE 4.18 SL process basics.

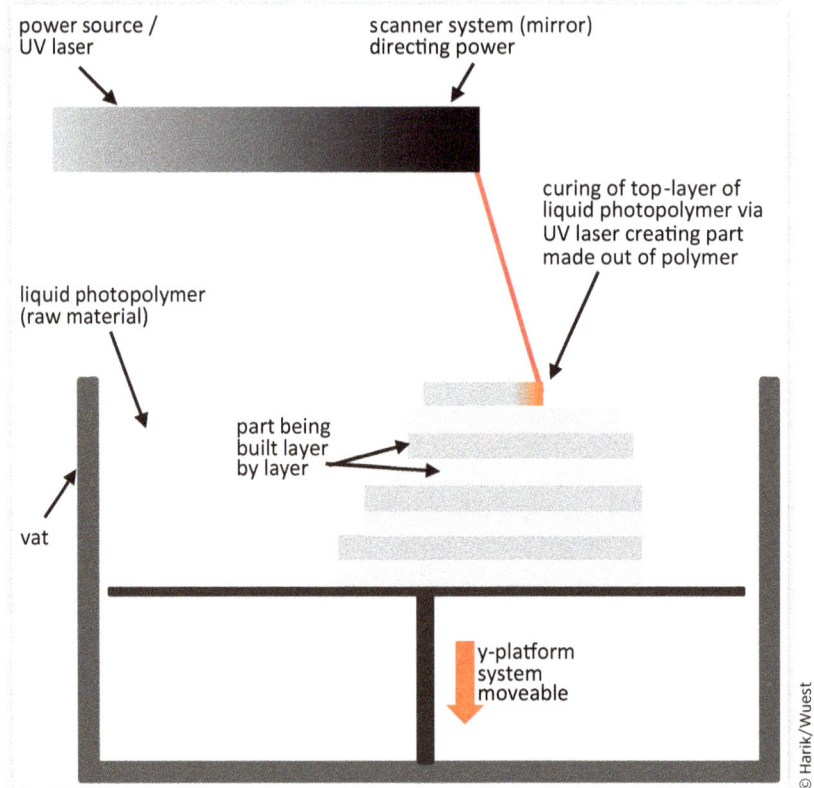

power source /
UV laser

scanner system (mirror)
directing power

curing of top-layer of
liquid photopolymer via
UV laser creating part
made out of polymer

liquid photopolymer
(raw material)

part being
built layer
by layer

vat

y-platform
system
moveable

© Harik/Wuest

FIGURE 4.19 SL tools.

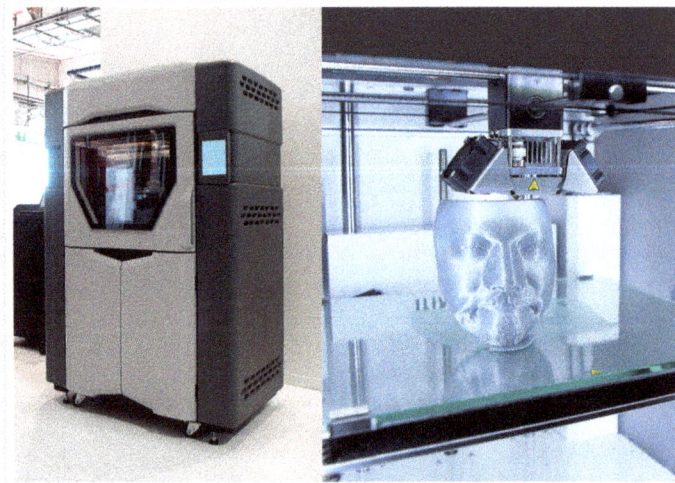

© Moreno Soppelsa/Shutterstock

require major initial investment and subsequent maintenance cost. The 3D Systems ProX 950 can go up to $1M while providing a large build area (59 × 30 × 22 in.) with an excellent 0.001 mm accuracy.

4.4.3 Applications

SL is currently widely spread due to multiple reasons: (1) high build quality, (2) multi-material printing, and (3) increased material variety. SL is used for visual prototypes, i.e., full-engine design prototypes; functional parts, i.e., industrial casting molds; and test prototypes such as medical applications and dentistry (Figure 4.20).

FIGURE 4.20 SL parts/products.

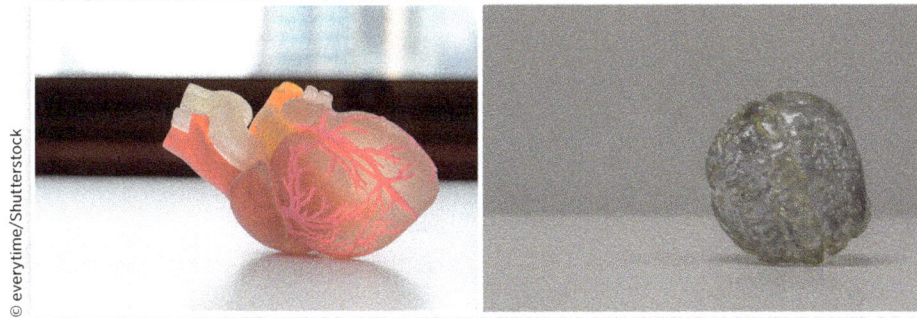

4.5 **Process Planning**

The below list represents the most common system part, found in AM equipment:

- **Fusion/consolidation**: Usage of lasers, electron beams, extrusion heads to create the localized solid entity

- **Cut/restart**: Usage of mechanisms and cutting knives that enable the cut/restart process

- **Pressure**: Usage of pressure mechanisms in certain processes to "sinter" components

- **Motion control**: Usage of mechanisms to ensure deposition/addition location and feedback to system

- **Containment chamber**: Usage of volume separation to contain heat/cold depending on process, i.e., cold chamber manufacturing

In its most basic form, AM modeling includes the Design, Tessellation, and Slicing steps (see Figure 4.21). Tessellation enables the conversion of the CAD model, resulting from the Design phase, into the STL (Standard Tessellation Language) format. Slicing creates the in-path plans for the layup procedure.

Process planning is the matchmaking step between design and manufacturing resources. It includes multiple functions such as Topology Optimization, Slicing and Toolpathing, Build Orientation, Grouping, and Nesting. The system functionality and the optimal processing is a function of a proper understanding of the system (3D Printer) parts, the modeling flow, and its effects.

4.5.1 **Slicing**

Slicing transforms the solid CAD model of the to-be-manufactured part to the set of individual layers that will ultimately make up the part and defines the toolpath (Figure 4.22). There are various software solutions available that aim to support the engineer during the slicing process and decision (Freeware: Cura and Slic3r). Multiple decisions that have to be taken during the slicing step have a direct impact on the final outcome of the part, including strength, surface quality, and tolerances. Some of these decisions are printer speed, fill density, retraction, shell thickness, and layer height/initial layer height. Highly dependent on the apparatus and material, some research is conducted in using build quality to secure digital

FIGURE 4.21 Model flow in AM.

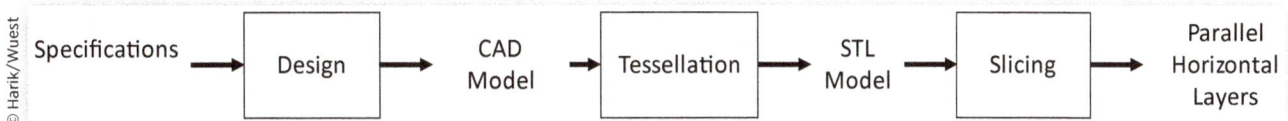

Specifications → Design → CAD Model → Tessellation → STL Model → Slicing → Parallel Horizontal Layers

FIGURE 4.22 Slicing algorithms (Halbritter et al., 2017).

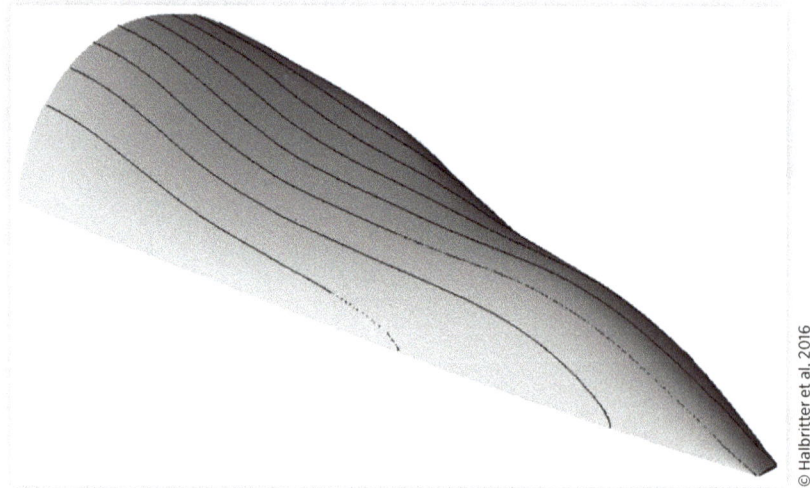

files and prevent counterfeiting. Moreover, there are several features that can cause problems during slicing, some of which will be discussed in more detail in the following AM challenges section.

4.5.2 Grouping/Nesting

Grouping describes the strategic arrangement of different (or multiple of the same) parts of an assembly to be printed in one setting (Figure 4.23). Well-placed parts during grouping can significantly enhance the output of the AM process and thus increase the economics significantly. Badly placed parts can not only "waste" valuable AM capacity but also cause built quality problems.

An example of such an arrangement that can cause quality issues is placing parts too closely together. Depending on the process (e.g., SLS), the dispersed energy can negatively effect the material property of too closely placed neighboring part.

FIGURE 4.23 Grouping/nesting (Zhang et al., 2016).

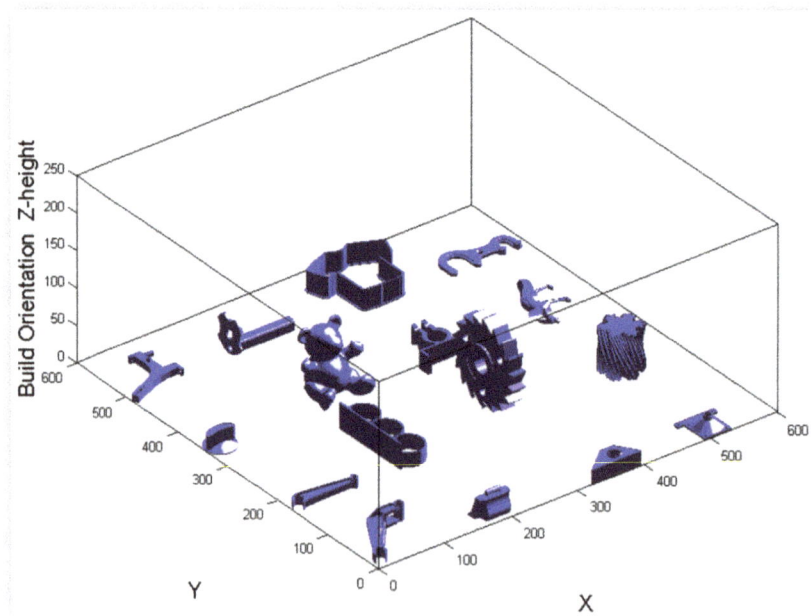

4.5.3 **Build Orientation**

Build orientation focuses on the virtual and physical placement of the part on the build platform (Figure 4.24). It is very important and can affect multiple parameters like build quality, material properties of the printed part, and even speed of manufacture. It has to be understood that slicing and build orientation are not independent but strongly depend on each other. Build orientation can have a significant impact on how much support material/structure is required to manufacture the part. An experienced engineer can save significant rework and material waste by strategically using build orientation to optimize the support needed for overhangs and such. As this directly reduces the need for rework and waste of (expensive) material, the economic impact of the build orientation strategy is immanent.

FIGURE 4.24 Optimal build orientation (Zhang et al., 2016).

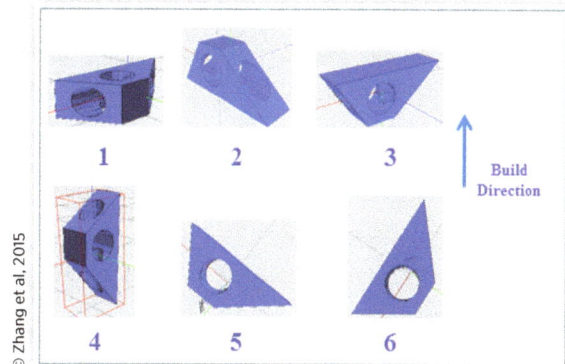

© Zhang et al, 2015

4.6 **Challenges of AM**

AM has multiple benefits and drawbacks. Subsequent efforts were conducted to automate the manufacturability analysis of AM, taking into account several of the previously identified challenges (Shi et al., 2018). Recent research demonstrates steady steps towards the implementation of an AM manufacturability tool. Currently, many professional guidelines are available for designers to make sure that the designs they generate are manufacturable and to avoid costly mistakes and/or extensive post-processing. These guidelines provide support for designers when it comes to multiple critical AM features such as (1) unsupported features, (2) minimum feature size, (3) maximum vertical aspect ratio, (4) minimum spacing, and (5) minimum self-supporting angle. It has to be noted, due to the very different nature and principles of the various AM processes, some challenges only apply to selected variations while others are generally applicable.

4.6.1 **Unsupported Feature**

FFF cannot extrude material over "thin air"; therefore this process requires **external support structures** for overhangs, bridges, and horizontal holes. Figure 4.25 depicts examples for these **three most common types of unsupported features**. The red arrows mark the decisive dimensions involved that the designer needs to take into consideration during her/his design decision. Some processes found an expensive solution to the unsupported feature challenge. It involved the manufacturing in a contained cold environment in a way that the material freezes without potential bending/failure issues. Some AM processes, other than FFF, do not necessarily require support structures when building these features. SLS, for example, has "built-in" support via the raw powder layers supporting the top layer naturally.

FIGURE 4.25 AM challenges: unsupported features.

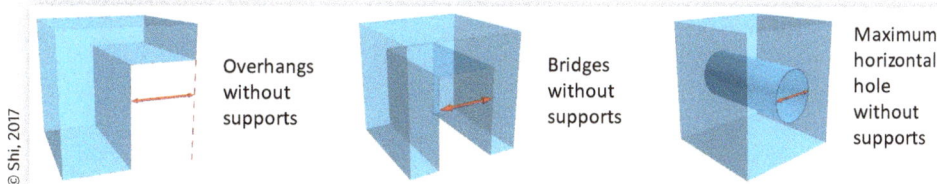

© Shi, 2017

Overhangs without supports

Bridges without supports

Maximum horizontal hole without supports

CHAPTER 4

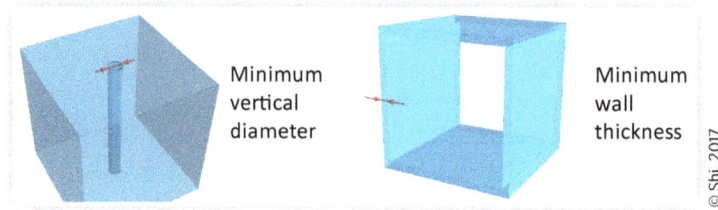

FIGURE 4.26 AM challenges: minimum feature size.

© Shi, 2017

4.6.2 Minimum Feature Size

In the AM process, thin-wall or small-size structures are subject to significant thermal dissipation (Figure 4.26). This thermal dissipation may cause various defects, such as unmelted powder inclusions, internal voids, cracks, and shape irregularities. Therefore, it is necessary to specify a minimum dimension for thin walls and holes. While this is true for basically all AM processes, the specific minimum wall thickness/feature size depends strongly on what process is chosen.

FIGURE 4.27 AM challenges: maximum vertical aspect ratio.

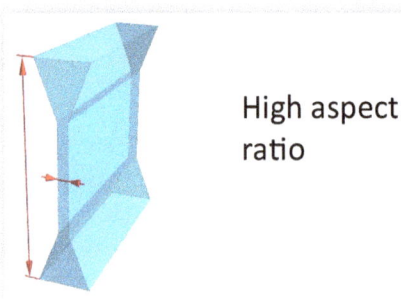

High aspect ratio

© Shi, 2017

4.6.3 Maximum Vertical Aspect Ratio

FFF feature cannot have a **vertical aspect ratio** exceeding a maximum value (Figure 4.27). The aspect ratio is defined as the proportional relationship between a feature's height and width. Continuation of the recoating process will eventually result in the **feature's bending**. In principle, this feature is similar to the "thin features" challenge faced in Subtractive Manufacturing.

4.6.4 Minimum Spacing

In Powder Bed Fusion (SLS/SLM) processes, if two **surfaces are too close** to each other, heat from one side may influence the properties of the other side (Figure 4.28). Therefore, it is necessary to specify a **minimum spacing** between two different surfaces.

FIGURE 4.28 *AM challenges: minimum spacing.*

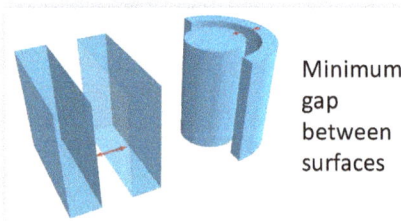

Minimum gap between surfaces

© Shi, 2017

4.6.5 Minimum Self-Supporting Angle

For FFF features, it is necessary to set a **minimum inclination angle** to ensure that angled faces will **not collapse** without support material (Figure 4.29). This can be understood as a variation of the "unsupported feature" challenge.

There is extensive research on AM conducted, notably (Shi, 2017), where Shi attempts to design an AM design support tool (Figure 4.30). Shi attempts to automate the process of identifying critical features (as depicted previously) to provide decision support for the designer. While the research led to interesting and promising results, there is no fully functional system available as of yet, and it still falls into the designers/production planners expertise to identify and analyze critical features before they cause problems during the AM process. These models were based of the Shapeterra software outlined in (Harik et al., 2017).

FIGURE 4.29 AM challenges: minimum self-supporting angle.

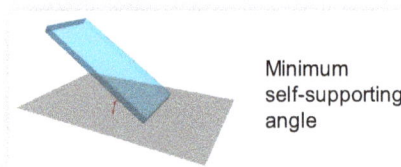

Minimum self-supporting angle

© Shi, 2017

FIGURE 4.30 Algorithms to detect AM challenges in an automated process (Shi, 2017).

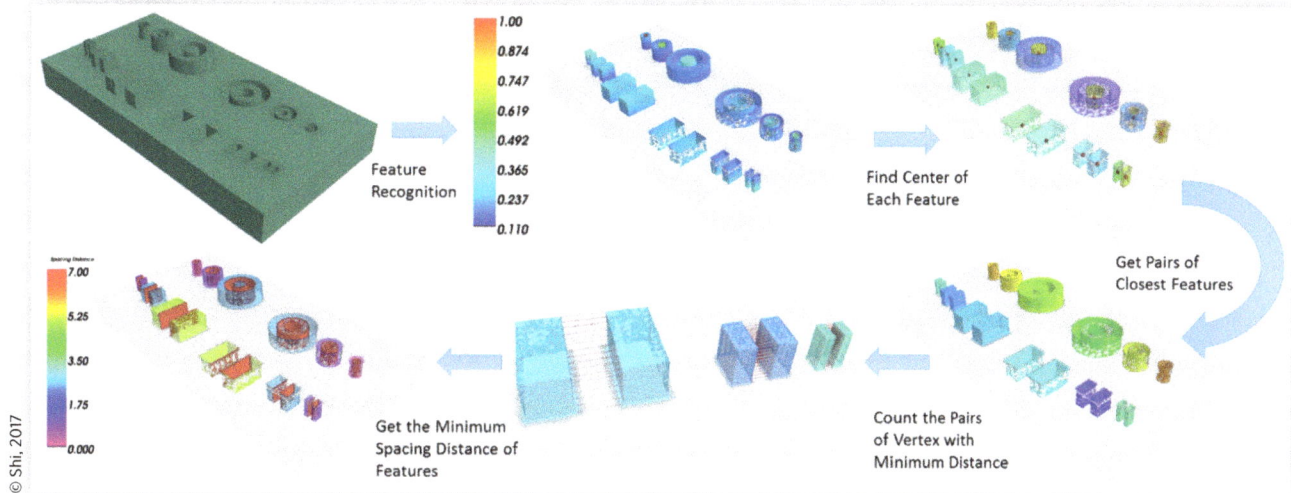

References

Halbritter, J., Harik, R., Zuloaga, A., van Tooren, M., "Tool Path Generation on Doubly-Curved Free-Form Surfaces," *Computer-Aided Design and Applications* 14, no. 6 (2017), 855–851.

Shi, Y., Zhang, Y., Baek, S., De Backer, W., Harik, R., "Manufacturability Analysis for Additive Manufacturing Using a Novel Feature Recognition Technique," *Computer Aided Design and Applications* 15, no. 6 (2018), 941–952.

Zhang, Y., Bernard, A., Gupta, R.K., and Harik, R., "Feature Based Building Orientation Optimization for Additive Manufacturing," *Rapid Prototyping Journal* 22, no. 2 (2016), 358–376.

Assembly Processes

In Chapters 2 through 4 we introduced the three main manufacturing families, Deformative, Subtractive, and Additive Manufacturing. In this chapter, we build on this foundation and present *Assembly Processes* as a means of combining two or more individual parts to a single product (or sub-product) (see Figure 5.1). Products that are made up of various parts are often referred to as assemblies. The different parts can be manufactured using a variety of different manufacturing processes, including those introduced in Chapters 2 through 4, 7, and 8. In this sense, Assembly Processes can be understood as tying the different manufacturing processes together, utilizing their individual strength and capabilities.

Assembly Processes are a very common technique used in a variety of ways across all industries. Historically, early humans have used Assembly Processes since they began to manufacture and use tools and weapons for hunting. Imagine a simple axe with a stone head, as depicted in Figure 5.2. At least two different parts had to be manufactured and assembled to create such a simple axe: a wooden shaft and a stone head. In order to securely attach the stone head to the wooden shaft, a means of joining has to be deployed – in this case, a rope made from either tendons or natural fiber was used to hold the two parts together.

Moving ahead to today's industrial age, assemblies are still omnipresent and can become extremely complex with millions of individual parts. One of these extreme examples is the Airbus A380 (see Figure 5.3a), an assembly with approximately four million individual parts sourced from 30 different countries and 1,500 suppliers.* Even a relatively common Mercedes-Benz SL500 (see Figure 5.3b) is assembled from approximately 9,100 individual parts that are joined using a variety of different techniques including over 5,000 spot welds. Overall, most of the products we use in our daily lives involve some form of Assembly Processes (and be it for the packaging they are delivered in).

Automotive companies such as General Motors, Ford, or BMW are notorious for reducing their vertical integration and are often specialized in design and assembly, while at the same time outsourcing most of the individual parts' production. This led to some large first-tier suppliers such as BOSCH, Johnson Controls, and ZF supplying parts to several different car companies. They often establish their own manufacturing locations in close proximity to the final assembly plants thus creating industry clusters.

---

* https://www.cnn.com/travel/article/airbus-a380-parts-together/index.html.

FIGURE 5.1 Product structure highlighting assembly processes.

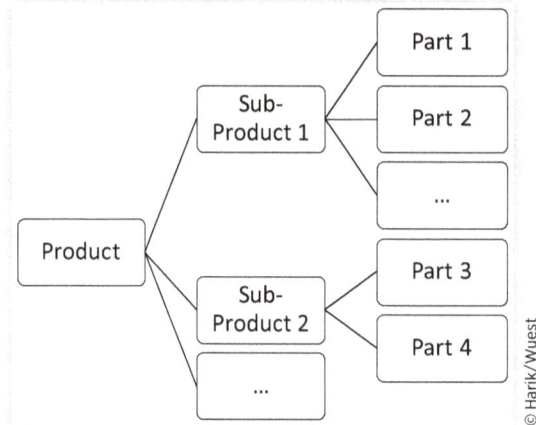

© Harik/Wuest

FIGURE 5.2 Simple Stone Age tools and weapons.

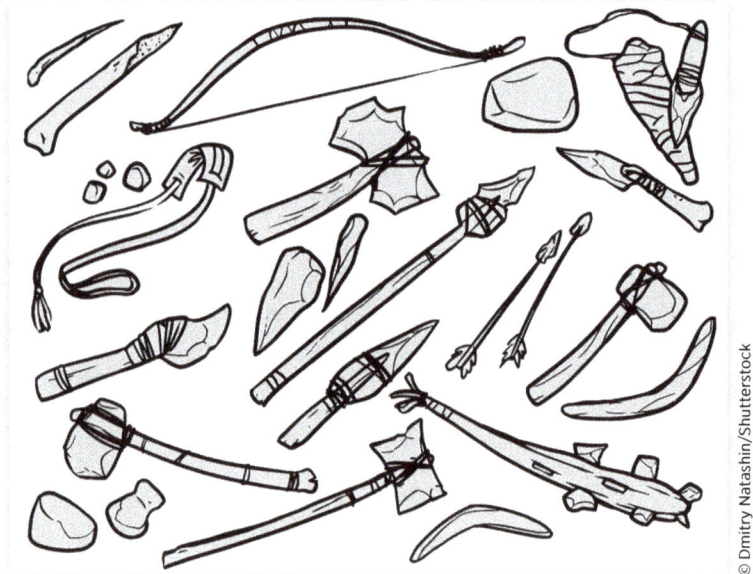

© Dmitry Natashin/Shutterstock

FIGURE 5.3 a) Airbus A380 superjumbo (left); b) Mercedes-Benz SL500 (right).

© Shutterstock

The remainder of this chapter is structured as follows: We first discuss the raison d'être of Assembly Processes and introduce a classification structure of the different available Assembly Process families. Then we have a closer look at permanent Assembly Processes in Chapter 5.2, before moving on to non-permanent Assembly Processes in Chapter 5.3. In Chapter 5.4, we cover some more exotic Assembly Processes to give a sense of the wide variation of technologies and techniques available in this field. Finally, we conclude the chapter with a section on Design for Assembly (DfA) where we take on the role of a designer and introduce some guiding principles for designing good assemblies.

5.1 **Classification of Assembly Processes**

Almost every minimally complex product is manufactured using some form of Assembly Processes. The reasons for designing a product as an assembly are manifold. Some are based on the bare necessity of materials used, others stem from economic and related (e.g., logistics) reasoning. It has to be noted that there is no clear division between technical and economic reasons at times. In most cases there are several reasons that apply at the same time and have to be considered simultaneously.

We structured the reasons based on the more prominent feature in those cases. In the following, we will discuss mainly technical reasons for the use of assemblies (Chapter 5.1.1) and the economic and other reasons for the use of assemblies (Chapter 5.1.2), and then introduce a classification structure of the main Assembly Process families and technologies in Chapter 5.1.3.

5.1.1 **Technical Reasons for Assemblies**

There are several different technical reasons for designing a product as an assembly. A general rule is that the more *complex a product* (product structure) the higher the likelihood that Assembly Processes are required. Take an automobile or an airplane with a variety of materials, electronic components, and highly stressed parts. It is not possible today to manufacture those complex systems without designing them as an assembly. At the same time, designing a product as an assembly does not necessarily mean you have to outsource the parts. As long as the resources are available, manufacturing parts for an assembly in-house might make sense, but should be critically reflected.

One of the most prominent technical reasons for opting for an assembly is when a product requires different (*contradicting*) *materials and/or varying material properties*. This can be an assembly with a metal hull that has to undergo heat treatment, but features integrated wooden elements. The heat treatment process is necessary to achieve the required properties of the metal part. However, it would destroy the wooden components through the high processing temperatures. Designing this system as an assembly allows to use the best (most suitable) manufacturing process for each component to achieve the best possible result and still have the final (assembled) product feature the wooden elements in the heat-treated metal hull. Another example is a regularly used kitchen item, a cooking pot. A cooking pan needs to have good conductivity to disperse heat in order for food to cook evenly. However, in order to pick up the pan the handle needs to be insulated ideally, e.g., using wood or a silicone polymer. Other common material combinations include, but are not limited to, glass/rubber, metal/glass, paper/metal, etc.

Another common reason for assemblies is the *availability of and/or access to required manufacturing resources and expertise*. If these resources and expertise are not available in-house at the manufacturer, the only option is to outsource the part manufacturing to a supplier that has the required resources and expertise. These resources can include, for example, a special CNC machine tool that can manufacture the required thin features or expertise in the area of manufacturing of certain exotic materials like graphene or nano/micro parts.

Similar to the previous point, another reason for designing a product as an assembly is the dimensions of the final product itself. In case the final product cannot be manufactured as an individual piece with the available manufacturing resources, e.g., due to the limited build area of certain computer numerical control (CNC) or selective laser sintering (SLS) machine tools, it might make sense to manufacture it in parts and assemble it later.

When the final product requires the *option to be disassembled* presents another reason for assemblies. The requirement for possible disassembly can stem from logistics requirements, such as offshore wind turbine (up to 180 m diameter) (see Figure 5.4a) where each blade (up to 88.4 m long each) already pushes the limits of transporting it on official roads with bridges and tunnels (see Figure 5.4b).

In cases where the final product design contains a large percentage of "void space" (empty space), assemblies should be considered as well. A prime example is a chair or table (see Figure 5.5a). This not only reduces the manufacturing complexity but also directly impacts the shipping and warehousing costs (see Section 5.1.2).

FIGURE 5.4 a) Off-shore wind turbine; b) Road transport of disassembled wind turbine blade.

© ER_09/Shutterstock; © rCarner/Shutterstock

FIGURE 5.5 a) Furniture fully assembled in showroom (left); b) warehouse with furniture in packages (right).

© Shutterstock

Another reason for assemblies is the possibility for easier *maintenance and repair*. When components can be easily removed and replaced, the product life cycle and remaining useful life of the product can be extended, and the initial investment is better utilized. An example for components that need to be regularly replaced are the brake pads in your car. This is increasingly true for upgradability of more complex systems as well. A prime example for life cycle extension through upgradability is the B52 strategic bomber. In service since the 1950s, it is expected to serve until 2050. Given the advances in communication technology and computing power, this would not be possible without upgrading the plane by replacing several parts including the jet engines.

A related reason for assemblies is when parts that are designed to "fail first" (*preferred failure mode*) allow a product to fail with minimal consequences for the overall system. Once the system fails, the part that is designed to fail first can be easily replaced and the overall system can be used again with minimum downtime and cost.

5.1.2 Economic and Related Reasons for Assemblies

One of the main reasons for choosing an assembly-based design is *logistics cost*. This includes but is not limited to shipping cost as well as storage (warehousing) cost. A good example for this is IKEA. Most of us have assembled a piece of IKEA furniture in our lives (see Figure 5.4a). IKEA is a prime example of how Assembly Processes can impact operations. IKEA's business model is largely built around the DIY (do it yourself) system that requires customers i) to pick out the disassembled and conveniently packed furniture directly from the IKEA warehouse (see Figure 5.4b) after experiencing it fully assembled in the showrooms, ii) transport the packages to their home, and iii) assemble the furniture following a (supposedly) user-friendly and well-documented Assembly Process with all materials (except some common tools) and instructions provided.

The question arises, why does IKEA's business model depend so strongly on the Assembly Process? Is it not counterintuitive to voluntarily omit a value-adding process that would allow charging a premium? By designing the Assembly Processes the way they do, they effectively leverage several advantages. They outsource the assembly operation to the customer which directly saves cost. Secondly, and most importantly, by having the furniture packaged in a well–thought-out way, minimizing the footprint, IKEA reduces shipping and storage cost and, therefore, can turnover significantly more products using smaller warehouses. This allows them to have their stores located in or close to densely populated areas, and thus in direct proximity to a significant number of potential customers.

Another economic reason for assemblies is the option to *outsource parts to suppliers* that can manufacture the part cheaper than the OEM would be able to by themselves. This is often associated with manufacturing in another country with all the associated parameters, such as taxes, labor cost, environmental laws, access to qualified workers, and cheap energy. This make-or-buy decision for whole products or parts is an important reason for assemblies.

We already discussed the technical reasons stemming from limitations of machine tools and tooling regarding the size of a product or workpiece. The other side of the medallion is the economic impact of required specialized or oversized tooling. Instead of designing a product with dimensions that require very expensive tooling or machinery, it might be more appropriate to design it as an assembly and use standard tooling/machinery. Overall it is often *more economical to assemble a large product* than to produce it in one piece.

There are many other reasons that make assemblies the design of choice. Some are not as obvious and based on common sense than the ones presented previously. One exotic example is *special import tax laws*. Some countries tax imports of fully assembled products very differently from products that require some finishing manufacturing steps. And these finishing manufacturing steps include assembly processes in most cases. As we know, manufacturing is considered a value adding activity and adds to a country's GDP besides providing jobs and purpose to the people. According to an executive of a subcontractor of Mercedes-Benz, the following scenario emerged when Mercedes-Benz wanted to import their G-Class to Russia prior to 2012*: The import taxes that would have been charged for a fully assembled G-Class were so significant that Mercedes-Benz decided to work with a subcontractor to first disassemble the fully functional new G-Class in Germany, package it for shipping, and then ship and import the disassembled G-Class to Russia, where they set up an assembly operation to reassemble the SUV on Russian soil. This, technically non-value adding activities were justified by significant lower import tariffs that made it a sensible and successful economic case leading to higher profits. This is an extreme example that showcases the various parameters that might play a role in decisions impacting the design, manufacturing, and assembly operations significantly.

Successful engineers today need to keep an open mind and use their skill set to adapt to the complex and dynamic environment to interpret these parameters in their design and manufacturing decisions.

5.1.3 Classification Structure of Assembly Processes

There are two major forms of joining assemblies: *permanent and non-permanent*. For both variations, there are distinct reasons for different applications. For some, being able to remove and replace a part is essential, so the only logical choice is a non-permanent assembly process. An example for such a case is the tire of a bike which has to be removable in order to change the mantel in case it is damaged or worn. In other cases, a part has to be permanently joined in an assembly. Reasons include safety and functional aspects. An example is the different layers of a smartphone display. And of course, there is the case where both options are feasible and the choice of permanent or non-permanent process is up to the designer.

In Figure 5.6 we illustrated a classification of common Assembly Processes. The overall distinction between permanent and non-permanent remains, and each is then subdivided in popular and widely used Assembly Processes. For *non-permanent assembly processes*, there are only fasteners, split again in threaded fasteners and pins, as well as some more exotic variations available, such as snap fits. For *permanent assembly processes*, we will discuss a wider variety.

* Russia became a member of the WTO in 2012 which lead to lower import tariffs.

FIGURE 5.6 Classification of assembly processes.

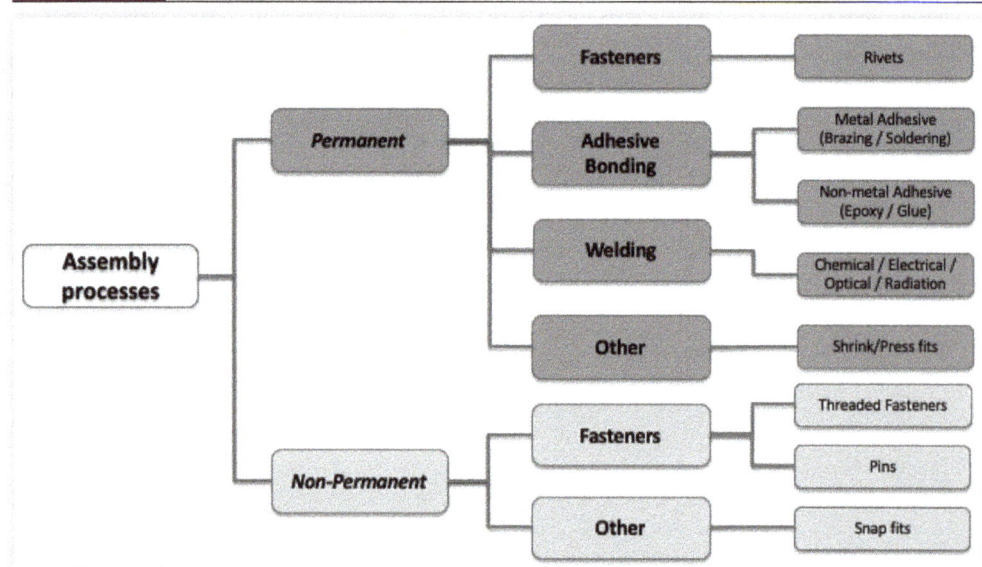

Three main subclasses can be distinguished: fasteners, adhesive bonding, and welding (also referred to as cohesive bonding). Permanent fastener types include rivets. Adhesive bonding is subdivided in metal adhesive bonding (brazing, braze welding, and soldering) as well as non-metal adhesive bonding (epoxies and glues). Welding is subdivided by type of welding process and by the energy source used. There are several other permanent Assembly Processes available, of which we will discuss shrink/press fits (including seams) in this chapter. All of these will be introduced in more detail in the following sections.

5.2 Permanent Assembly Processes

Permanent Assembly Processes describe tools and methods that are used to join two or more parts of an assembly without the option to disassemble and taking them apart without destructive measures. It is often the case that there is a mix of permanent and non-permanent Assembly Processes used within the same assembly. This can be achieved in different variations. An example is when different parts are permanently (e.g., gluing a laminate on a wooden bar) and non-permanently (e.g., using a screw to attach the laminated wooden bars to a table as legs) joined to create a dining room table. In some cases, generally non-permanent assembly methods are made permanent by mixing assembly processes. Examples are welding a stud to a part, gluing a pin in a board, or welding a nut on a bolt.

In this subsection, the most common permanent Assembly Processes are presented. We follow the classification depicted in Figure 5.5 and start with permanent fasteners, focusing on rivets, (metal and non-metal) adhesive bonding, and finally the different welding processes available. Shrink/press fits, while considered a permanent assembly process, are presented in Section 5.4 under recent trends and non-traditional techniques.

5.2.1 Permanent Fasteners/Rivets

When talking about permanent fasteners, in most cases the referred to process is riveting. Rivets are a very common type of permanent fastener used in a variety of industries. Compared to threaded fasteners rivets are considered more economical, stronger, and faster to apply. Rivets are designed to addressing both shear and tension loads. However, there is a sheer endless number of different fasteners available, including several permanent variations. Many of those are developed for very specific applications; therefore, we consider them out of scope for this chapter as we aim to provide an overview of commonly used tools and methods.

There are two types of rivet classes, i) *solid rivets* which are made of solid material and used when both sides of an assembly can be accessed. The other variation, the so-called ii) *blind rivets*,

FIGURE 5.7 Blind rivets and rivet gun.

© Alberto Masnovo/Shutterstock

are made from hollow material and used when only one side of the assembly is accessible. It does not come as a surprise that solid rivets are stronger when the material and weight of the rivet are comparable. Hence, to achieve similar performance, blind rivets need to either i) use stronger material and/or ii) be dimensioned larger, and/or iii) be used in greater numbers. In practice, blind rivets are often heavier than solid rivets due to the use of steel stems (mandrel) (see Figure 5.7).

Several products commonly using rivets, including airplanes (see Figure 5.8a), bridges (see Figure 5.8b), jeans (see Figure 5.8c), and boats (see Figure 5.8d). Rivets are also used regularly in building construction and most home appliances and many more other applications we use on a daily basis. To provide a rough orientation on the numbers of rivets used for common products, as an example, the Boeing 747-8 airplane uses over 1,000,000* rivets in its assembly and the Airbus A380 750,000† rivets.

FIGURE 5.8 a) Rivets on a steel bridge (top-left); b) Rivets on an airplane (top-right); c) Rivets on a classic jeans (bottom-left); d) Rivets on a boat hull (bottom-right).

© Dan Henson/Shutterstock; © DenisProduction.com/Shutterstock; © Shutterstock; © Marbury/Shutterstock

---

* http://www.aeronewstv.com/en/lifestyle/in-your-opinion/2782-a-boeing-747-8-has-how-many-rivets.html.
† https://www.theguardian.com/business/2006/feb/23/theairlineindustry.travelnews.

FIGURE 5.9 Blind rivet process steps 1) to 4).

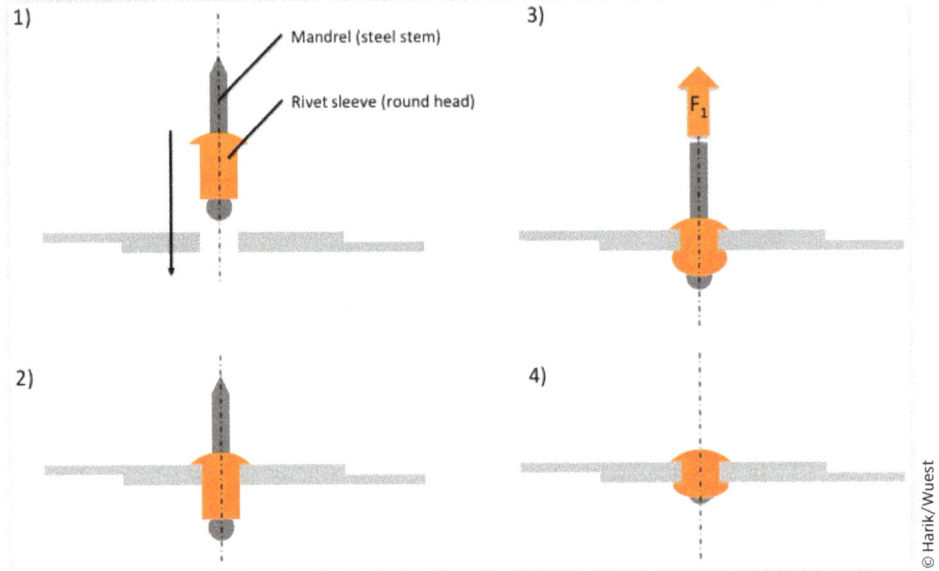

The process of inserting and fastening a blind rivet can be divided into four distinct steps (see Figure 5.9). In the first step (1), a hole is drilled or punched in two adjacent sheets of material (most commonly sheet metal). Then (2) the blind rivet is inserted in the hole. Once the head is aligned with the sheet metal, (3) the rivet gun pulls the internal mandrel and thus deforms the rivet sleeve to effectively join the two sheets together. Once the force exceeds a certain amount, (4) the mandrel breaks (intentionally) and the rivet is installed, securely holding the two sheets permanently together.

5.2.2 Adhesive Bonding

Adhesive Bonding processes describe the joining of two or more workpieces by using an adhesive as a bonding agent between the two or more workpiece materials being joined.

There are two main categories of Adhesive Bonding processes: Metal Adhesive Bonding and Non-metal Adhesive Bonding. Metal Adhesive Bonding includes Soldering and Brazing (including Braze Welding), while Non-metal Adhesive Bonding includes Glues and Epoxies.

5.2.2.1 METAL ADHESIVE BONDING

The two main forms of metal adhesive bonging are soldering and brazing (including Braze Welding). Both Soldering and Brazing (including Braze Welding) are described as adhesive bonding process using a metallic filler material as the bonding agent.

Soldering (see Figure 5.10) distinguishes itself from brazing (and Braze Welding) by the metallic filler material used. In soldering, the metallic filler material has a liquidus below 840°F (450°C) and below the solidus of the base metal. The metallic filler material is distributed by capillary attraction during the soldering process. Soldering metallic filler material typically contains metals such as lead, tin, and cadmium.

Brazing on the other hand, while also describing an adhesive bonding process using a metallic filler material as the bonding agent, uses a metallic filler material that has a liquidus above 840°F (450°C) and below the solidus of the base metal (see Figure 5.11). Similarly, the metallic filler material is distributed by capillary attraction during the brazing process. Brazing metallic filler material typically contains metals such as copper, silver, and gold. Braze Welding is a variation of brazing that does not use capillary attraction to disperse bonding metal material. Generally, brazing is used on joints with a larger contact area while braze welding is used for fillet-type joints.

5.2.2.2 NON-METAL ADHESIVE BONDING

There are two commonly used forms of non-metal adhesive used in industry: Glues and Epoxies. Glues are technically also epoxies, referred to as one/single-component epoxies.

FIGURE 5.10 Soldering process on a circuit board.

© tcsaba/Shutterstock

FIGURE 5.11 Brazing process to join several parts of a copper tube assembly.

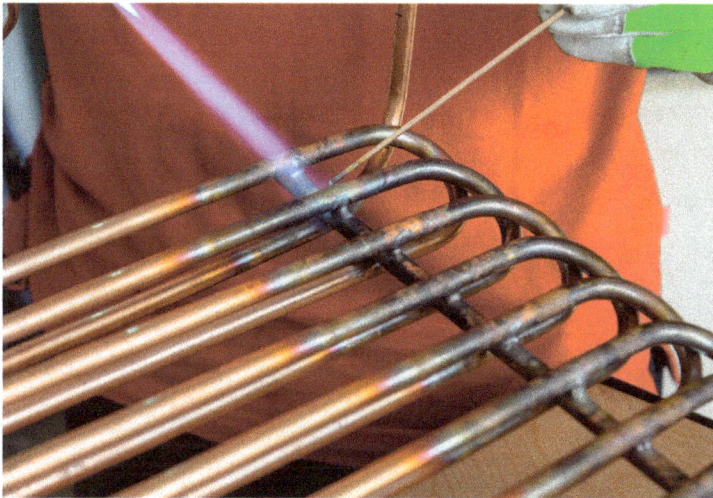

© Bildagentur Zoonar GmbH/Shutterstock

FIGURE 5.12 Two-component epoxy applicator.

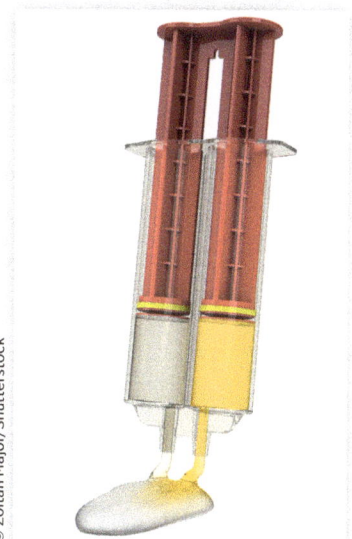

© Zoltan Major/Shutterstock

They are different from regular, two-component epoxies in the sense that adhesives that they use heat as the activating curing (hardening) agent. (Two-component) Epoxies on the other hand contain two different components, both the resin and curing (hardening) agent (see Figure 5.12). The reaction that allows the epoxy to perform as an adhesive is activated when the two components are mixed together. Two-component epoxies are the most commonly used variant in industrial application. Other forms of adhesives include silicones, acrylics, and several other specialized ones.

The processes of applying non-metal adhesives can be generalized as follows: 1) prepare surface(s), 2) prepare epoxy or glue, 3) apply epoxy or glue, 4) apply pressure, and 5) cure/harden.

Epoxies and glues as adhesive bonding agents are often considered structural adhesives that allow the joining of dissimilar materials (e.g., metal/wood/glass). Furthermore, they can join materials with little or no elevated temperature which is required for some applications (e.g., very thin materials). In additional to the purpose of joining two or more parts of an assembly, epoxies and glues can fulfill additional design purposes, such as a sealant and insulation. Overall, epoxies and glues are (generally) economical and easy to apply compared to other adhesive and fastener assembly processes. However, they have several downsides that need to be considered when planning an assembly with the use of epoxies or glues. These include but are not limited to their comparably lower joint and peel strength, their restriction to lower temperature applications, their low conductivity, and their tendency to negatively impact production time due to a relatively long curing time.

Adhesives can be classified by different parameters like viscosity, curing speed, applicable temperature range, matching with to-be-joined material(s), and strength.

DID YOU KNOW?

In recent years some interesting innovations in this space emerged. A fascinating one is the use of bionic (bio-inspired) principles of the strong natural adhesive used by "blue mussels" (*Mytilus edulis*) to attach to underwater structures, like hulls of large shipping vessels. An artificial version of the natural adhesive is very desirable for a variety of applications, including medical and dental, which is superstrong and can be applied underwater (Wiegemann, 2005). Underwater joining is generally a very complex and challenging endeavor that complicates the use of other common assembly processes like traditional adhesives or welding.

Materials used as adhesives are mostly (natural and synthetic) polymers, but there are also some selected applications for ceramic adhesives (not that common). The polymers used as adhesives range from thermosets, thermoplastics, and elastomers. The original states of adhesives can take liquid, gel, paste, or solid form. However, liquid, paste, and gel types are the most common variations – and generally preferred due to a more convenient application process on the material surfaces (Figure 5.13).

Adhesive bonding processes require some design considerations for the joints. Typical joint design for adhesive bonding is the lap joint (see different variations depicted in Figure 5.14a-d).

FIGURE 5.13 Rubber-based adhesive applied to join two parts of a shoe assembly.

© Nor Gal/Shutterstock

FIGURE 5.14 Common lap joint variations: a) Single lap; b) Double lap; c) Bevelled lap; d) Double butt lap.

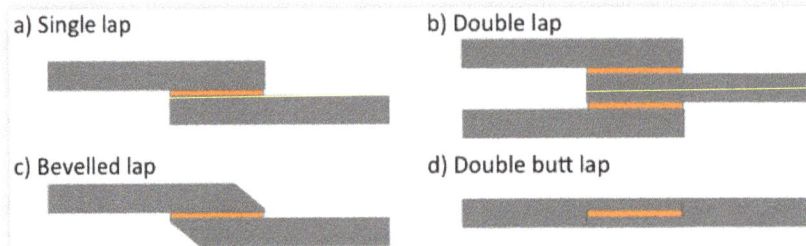

a) Single lap
b) Double lap
c) Bevelled lap
d) Double butt lap

© Harik/Wuest

The typical loading that the joint is designed and dimensioned for is shear stress to prevent a failure of the joint. Other types of stress that should be considered are tensile, cleavage, and peel stress.

5.2.3 Welding

Welding is a widely deployed permanent Assembly Process and is often shown in movies and pictures when a manufacturing environment is presented. The iconic welding mask, bright flame, and rough metal make for a great background (see Figure 5.15). A welding process describes the permanent joining between two (metal) workpieces and/or a workpiece and a metallic filler material.

In order to achieve the desired properties of the weld, proper *welding joint* design is essential. Selecting the best/most suitable joint design for the purpose depends on a variety of factors, such as the base metal material (carbon vs. stainless steel), the filler/weld metal material, and the employed welding technique (e.g., gas tungsten arc welding vs. shielded metal arc welding). Figure 5.16a illustrates a selection of commonly used welding joint variations, and Figure 5.16b presents additional joint terminology focused on the relative arrangement of the to-be-joined parts.

There are several different *welding types* used to achieve the desired joining of the two metal parts. Two types are particularly common, fillet welds and groove welds. Fillet welds describe the joining of two or more workpieces at a ~90° angle (see Figure 5.17b). The type can be further specified, such as a full fillet that is defined as a fillet weld where the size of weld equals the thickness of the smaller workpiece. Groove welds are similarly common to fillet welds and describe the welding bead joining two or more workpieces (see Figure 5.17a). For beveled groove

FIGURE 5.15   Welding process.

© Sasin Tipchai/Shutterstock

FIGURE 5.16   a) Selected commonly used welding joint designs; b) Joint terminology based on part arrangement.

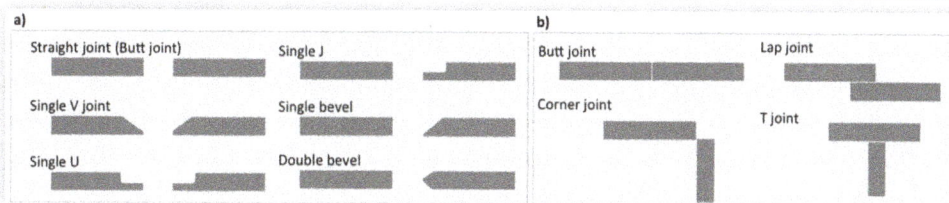

© Harik/Wuest

FIGURE 5.17 a) Different groove weld types; b) Fillet weld type.

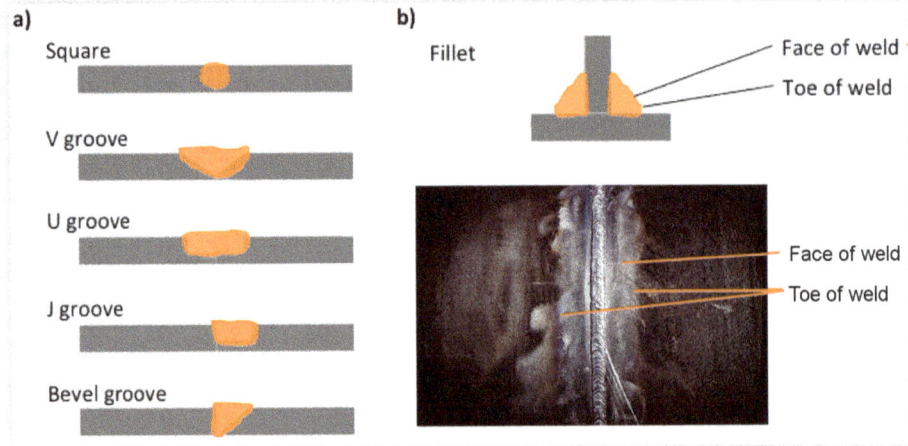

© Harik/Wuest; © Harik/Wuest; © Roman023_photography/Shutterstock

weld types, common angles are 10°, 15°, 22.5°, 32.5°, and 45°. There are several other less commonly used specialty variations of welding types available, such as surface welds, seam welds, flash welds, spot welds, and upset welds.

Depending on the type of weld, we can calculate the weld volume and weight when we know the area (A) and the length of the weld (l_{weld}) seam (see Figure 5.18). We can calculate the volume of the weld like this:

$$V = A * l_{weld}$$

Eq. (5.1)

FIGURE 5.18 Surface area of fillet and V groove welds.

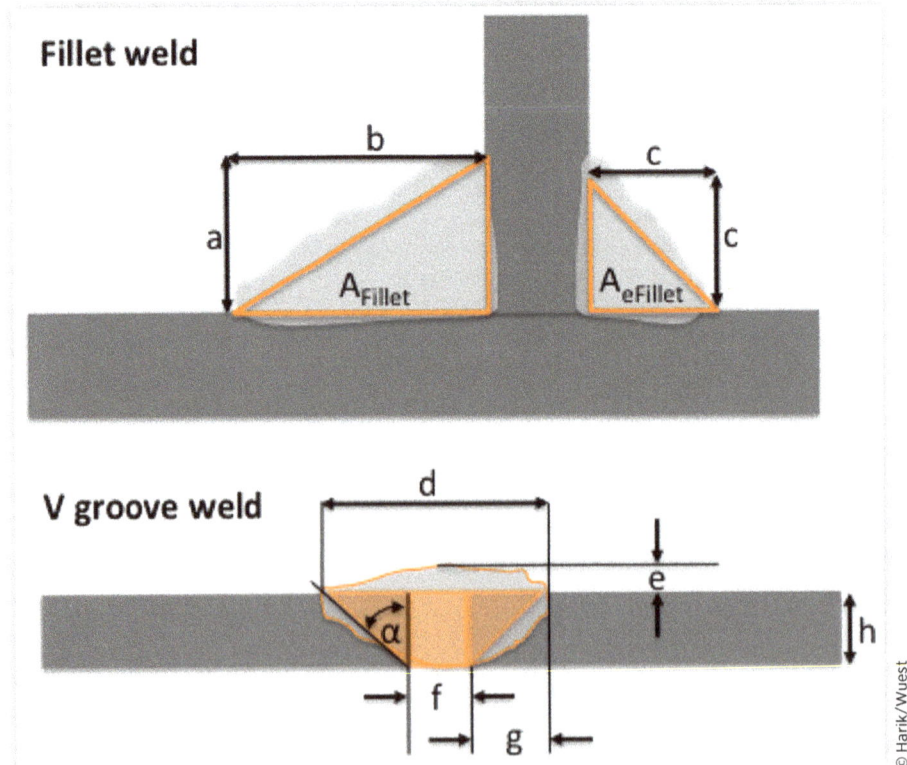

© Harik/Wuest

and the weight by multiplying with the material density of the weld metal:

$$W = V * d$$ Eq. (5.2)

It has to be noted that the weld is often contaminated with other materials; therefore the calculation is providing no exact values but a more or less accurate weight of the weld.

For fillet type welds, we have to distinguish if the weld has equal "legs." The legs end in the "toe" of the weld as depicted in Figure 5.17b. For fillet welds with equal legs ($A_{eFillet}$), we can easily calculate the area following this equation (see Figure 5.18 top-right):

$$A_{eFillet} = c^2/2$$ Eq. (5.3)

For fillet welds with unequal legs, as illustrated in Figure 5.18 top-left, we can determine the area (A_{Fillet}) in a similar fashion as we still deal with a right triangle:

$$A_{Fillet} = (a*b)/2$$ Eq. (5.4)

In case we need to use a butt weld with a single V groove (Figure 5.18 bottom) and want to know the area (A_{sVB}) of the weld, we have to use trigonometry. In order to determine A, we split the area of the weld into three to four distinct areas: 1) the gap that is filled with weld metal (root gap) (A_{sVB1}), 2 and 3) the two triangles of the bevel (A_{sVB3} and A_{sVB3}), and 4) the excess material building above the weld (A_{sVB4}).

$$A_{sVB} = A_{sVB1} + A_{sVB2} + A_{sVB3} + A_{sVB4}$$ Eq. (5.5)

Depending on what parameters are available, we can calculate the area. For the rectangular area of the root gap (A_{sVB1}), we simply use this equation:

$$A_{sVB1} = f*h$$ Eq. (5.6)

For the two bevel triangular areas, we can calculate the areas (A_{sVB2} and A_{sVB3}) as such:

$$A_{sVB2} = (g*h)/2 = (h*(h*\tan\alpha))/2$$
$$A_{sVB3} = (g*h)/2 = (h*(h*\tan\alpha))/2$$ Eq. (5.7)

We can simplify and calculate both areas in one equation:

$$A_{sVB2\&3} = (g*h) = (h*(h*\tan\alpha))$$ Eq. (5.8)

Different from the three areas calculated thus far, for excess material built up on the face of the weld, we use an approximation:

$$A_{sVB4} = (d*h)/2 = ((2(h*\tan\alpha)+f)*e)/2$$ Eq. (5.9)

Example Calculation

We are building a metal plate with a final length of 60.25 in. as a working surface by welding together two 1 in. × 10 in. × 30 in. plates. We choose a single V butt weld to join the two plates. The bevel angle α = 22.5° and the weld excess cap has a height of 0.1 in. *What is the volume of the weld?*

First, we need to calculate the area of the weld:

$$A_{sVB} = A_{sVB1} + A_{sVB2} + A_{sVB3} + A_{sVB4}$$

Eq. (5.10)

We know the plates that are joined are 1 in. thick and the final length of the table plate is 60.25 in. With the individual plates being 30 in. long, the root gap remains to be 0.25 in.

$$A_{sVB1} = f*h = 1 \text{ in.} * 0.25 \text{ in.} = 0.25 \text{ in.}^2$$

Eq. (5.11)

For the two triangular areas, we know that the bevel angle is the same for both plates (22.5°) and the plates are 1 in. thick.

$$A_{sVB2\&3} = \left(h*(h*\tan\alpha)\right) = \left(1 \text{ in.} * (1 \text{ in.} * \tan 22.5°)\right) = 0.558 \text{ in.}^2 \quad \text{Eq. (5.12)}$$

For the excess material buildup of the weld cap, we approximate the area:

$$A_{sVB4} = (d*h)/2 = \left((2*(h*\tan\alpha)+f)*e\right)/2$$
$$= \left((2*(1 \text{ in.} * 0.558)+0.25 \text{ in.})*0.1 \text{ in.}\right)/2$$
$$= 0.068 \text{ in.}^2$$

Eq. (5.13)

Therefore, the complete area is

$$A_{sVB} = 0.25 \text{ in.}^2 + 0.558 \text{ in.}^2 + 0.068 \text{ in.}^2 = 0.876 \text{ in.}^2$$

Eq. (5.14)

The length of the weld is equal to the width of the plate, which is provided to be 10 in. Hence, we can calculate the weld volume to be

$$V = A_{sVB} * l_{weld} = 0.876 \text{ in.}^2 * 10 \text{ in.} = 8.76 \text{ in.}^3$$

Eq. (5.15)

Welding processes employ typically high temperatures that are at or near the material melting temperature. The operating temperatures are thus typically higher than that of lower temperature techniques such as brazing/soldering described in the previous subsection. Typical operating temperatures for different processes are depicted in Table 5.2.

There are several different categories and designations of existing welding processes. Table 5.1 illustrates selected common welding classifications following the American Welding Society's (AWS) designations. The classification categories are determined based on the mode of energy transfer. The welding processes that we discuss in more detail in this subsection are highlighted with a light gray backdrop. We selected the welding processes based on their frequency of occurrence and combined coverage of the field.

For several of the welding processes, an *electrode* is a key system component. There is a large number of different electrodes available that can be divided in ones with consumable (e.g., SMAW, SAW) and non-consumable (e.g., GTAW, PAW) electrodes. In case of consumable

TABLE 5.1 Selected welding process designations according to American Welding Society (AWS)

Category	AWS classification	Abbreviation
Oxyfuel gas welding (*torch welding*)	Oxyacetylene welding	OAW
	Oxyhydrogen welding	OHW
	Pressure gas welding	PGW
Arc welding	Carbon arc welding	CAW
	Shielded metal arc welding	SMAW
	Gas metal arc welding	GMAW
	Gas tungsten arc welding	GTAW
	Flux-cored arc welding	FCAW
	Submerged arc welding	SAW
	Plasma arc welding	PAW
	Stud welding	SW
Resistance welding	Resistance spot welding	RSW
	Resistance seam welding	RSW
	Projection welding	RPW
Solid-state welding	Forge welding	FOW
	Cold welding	CW
	Friction (stir) welding	FRW
	Ultrasonic welding	USW
	Explosion welding	EXW
	Roll welding	ROW
Other/special welding Processes	Thermite welding	TW
	Laser beam welding	LBW
	Electroslag welding	ESW
	Flash welding	FW
	Induction welding	IW
	Electron-beam welding	EBW

© Harik/Wuest based on AWS

TABLE 5.2 Typical welding energy sources and their achievable parameters

Energy source	Welding process	Power density [W/cm^2]	Operating temp. [°C]
Chemical	Oxyacetylene welding (OAW)	$<10^3$	2,000-4,000
Electrical	Shielded metal arc welding (SMAW)	10^4	6,000-8,000
	Plasma arc welding (PAW)	10^6	15,000-30,000
Optical/radiation	Electron-beam welding (EBW)	10^7	20,000-25,000
	Laser beam welding (LBW)	$>10^8$	>30,000

© Harik/Wuest

electrodes, the electrode material is included in the weld pool, which can be desired or not depending on the process. Another differentiation for electrode types is continuous and noncontinuous (stick) electrode. Continuous electrodes are often also classified as consumable electrodes and are fed automatically through the weld gun into the weld pool.

AWS also provides a standard numbering system for stick electrodes. The system is similar to the SAE steel designations we introduced in Chapter 1. The system employs a combination of a single letter and four or five digits to describe a certain stick electrode. The letter defines the type of welding process the electrode is designed for, such as "E" for Arc Welding. The first two digits (or three for a 5-digit numbers, respectively) inform about the minimum tensile strength, e.g., E60XX for a 60,000 psi tensile strength electrode. The second to the last digit defines the welding position(s) the electrode is certified for. A "1" in this case describes an electrode that is certified for all positions, while a "2" can only be used for horizontal positions. The last digit defines the type of coating and the welding current that can be used. For example, an electrode with the designation E6013 describes an electrode for Arc Welding that can be used in all positions and is generally used with an AC current.

FIGURE 5.19 Gas welding process schematic.

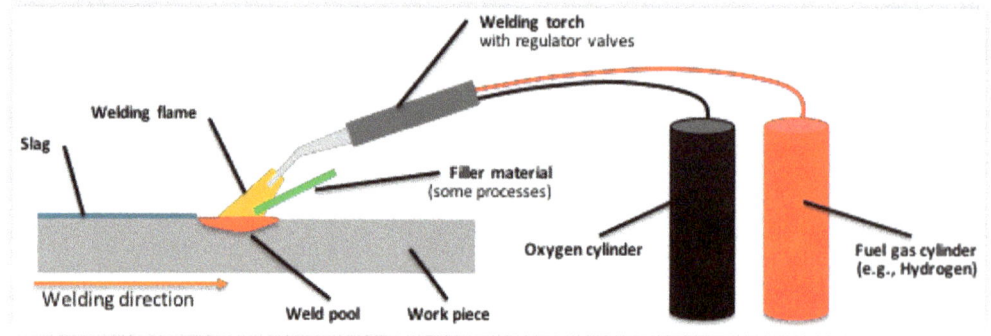

What it comes down to is delivering the *energy for the welding process* that allows for the fusion of the two metals (and the filler metal depending on the process), ideally in a precise way without negatively impacting the material properties. The energy can be delivered by different energy sources, including *electrical* (e.g., Arc Welding, Resistance Welding), *mechanical* (e.g., Friction Welding, Forge Welding), *chemical* (e.g., Thermite Welding, Gas Welding), and *optical/radiation* (e.g., Electron-beam Welding, Laser Beam Welding). A general rule is the higher the energy source the deeper the penetration. Laser beam welding is an example of a high-energy source that can achieve deep penetration of 1+ inch. Generally possible maximum power densities delivered by the various energy sources, not including mechanical, and the achievable operating temperature ranges are depicted in Table 5.2.

In the following we have a closer look at five typical welding processes from each group: Oxyacetylene Welding (Gas or Torch Welding), Gas Tungsten Arc Welding (Arc Welding), Resistance Welding, Forge Welding (Solid-State Welding), and Laser Beam Welding (Special Welding Processes). We will look at Induction and Friction Stir Welding as not so commonly used but very interesting welding processes in Subsection 5.4.

Oxyacetylene Welding (OAW) is a typical Gas Welding (often also called *Torch Welding*) process that is based on the principle of fusing two or more workpieces together using thermal energy (flame), often involving a filler metal. Torch welding has the advantage that the energy induced as well as amount of added filler material can be closely controlled by the (experienced) operator (Figures 5.19 and 5.20).

Arc Welding is one of the most commonly used welding processes in industry.

Arc welding induced the energy required to fuse two or more metal workpieces together using an electric power source. The Arc Welding process creates an arc between electrode and

FIGURE 5.20 Gas welding process on steel part.

FIGURE 5.21 Arc welding process schematic.

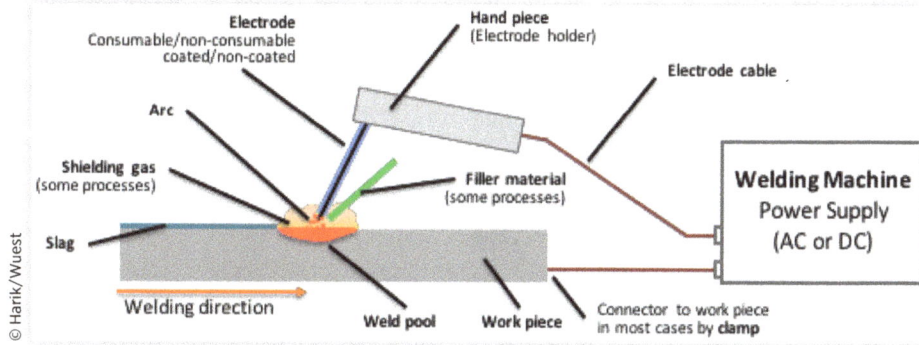

FIGURE 5.22 Arc welding process.

workpiece(s), effectively melting the material preparing the subsequent fusion. The melting process in this case is mostly affecting the base metal but with a consumable electrode, the electrode metal is included in the weld pool (and later the weld) as well (Figures 5.21 and 5.22).

Resistance Spot Welding (RSW) (often simply referred to as Spot Welding) uses electricity and pressure from both sides to create a welding nugget to join two (sheet) metal parts (see Figure 5.23a). Resistance spot welding is a very common process in the automotive industry especially in sheet metal joining processes around the cars' body. Resistance Spot Welding in the automotive industry is highly automated and integrated with robotic platforms allowing for high, repeatable quality and efficiency. The resistance spot welding cycle includes the application of current for a short interval while maintaining the joint under pressure (see Figure 5.23b).

FIGURE 5.23 a) Resistance spot welding equipment (left); b) Typical resistance spot welding process sequence (right).

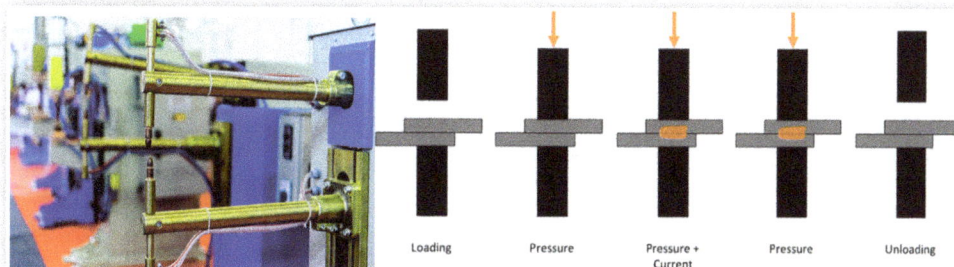

© Emituu/Shutterstock; © Harik/Wuest

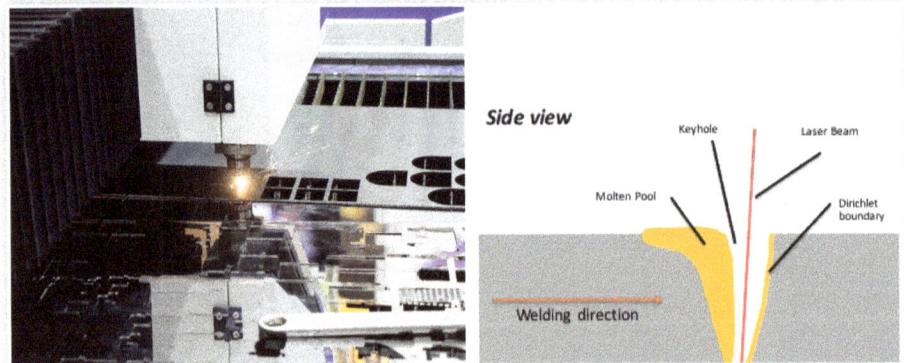

Forge Welding (FOW) is a solid-state welding process that has been used for centuries. In the forge welding process, the coalescence of the two metal materials is created by mechanical pressure and additional heat (depending on the process). A variation of forge welding, Friction Stir Welding is discussed in the following Chapter 5.4. Forge welding can occur in a metal extrusion process (see Chapter 3), when the remaining butt of the original billet is weld together with the next billet by the pressure of the ram before the material is extruded through the die. This allows an almost continuous extrusion similar to the truly continuous extrusion we experience in polymers manufacturing (see Chapter 7). However, the weld seam is often visible even in the extruded shape.

Laser Beam Welding (LBW) as a relatively new welding process describes the process of fusing two or more workpieces by inducing energy via a guided laser system. Laser beam welding combines two desired properties: high speed and high precision. As it delivers the energy very precisely and concentrated, the (thermal) impact on the rest of the part is reduced compared to other processes like Arc Welding. The general Laser Beam Welding process works as follows: The laser vaporizes metal material where it meets the base metal and creates a so-called keyhole which remains open as long as the process is running. The keyhole is then subsequently filled by the molten material of the weld pool (see Figure 5.24). The Dirichlet Boundary, which is the initial contact area of the laser with the material in welding direction, has a fixed temperature. Above the keyhole we often experience vapor or splatter similar to other welding processes.

The process is very complex and thus difficult to control. The high speeds achievable using a Laser Beam Welding(LWB) process add to the challenges for closed-loop control. While Laser Beam Welding speeds can reach more than up to 1200 mm/s (Dilip et al., 2016), generally high-speed Laser Beam Welding is associated today with speeds of 50-200 mm/s in order to achieve the required penetration and weld quality. Using high-speed cameras as an input for process control challenges even the most advanced analytics methods due to the enormous amount of data produced. For a comparable process (Selective Laser Melting), the amount of data generated is reported as up to 75 gigabytes of data every second (Everton et al., 2016). This is beyond the capabilities of most IT/OT systems and does (today) not accommodate real-time control. However, there are camper and machine learning-based approaches available that provide quality monitoring support for laser beam welding (Figure 5.25; Wang et al., 2017).

5.2.3.1 WELDING PHYSICS

In welding, the most important aspect of the process is the heat transfer that enables the joining of the two metals by coalescence. Figure 5.26 depicts the physics behind the heat transfer during a generic welding process. Depending on the process the details may vary (e.g., Laser vs. Gas) but the principles behind the heat transfer remains the same.

The amount of heat performing the welding operation experiences two losses from the source (H_i), the loss due to heat transfer (f_1) and the loss due to conduction (f_2). This allows us to formulate the following equation, describing the heat energy ($H_{welding}$) available for the actual welding process.

$$H_{welding} = f_2 f_1 H_i \qquad \text{Eq. (5.16)}$$

FIGURE 5.25   Laser beam welding process schematic.

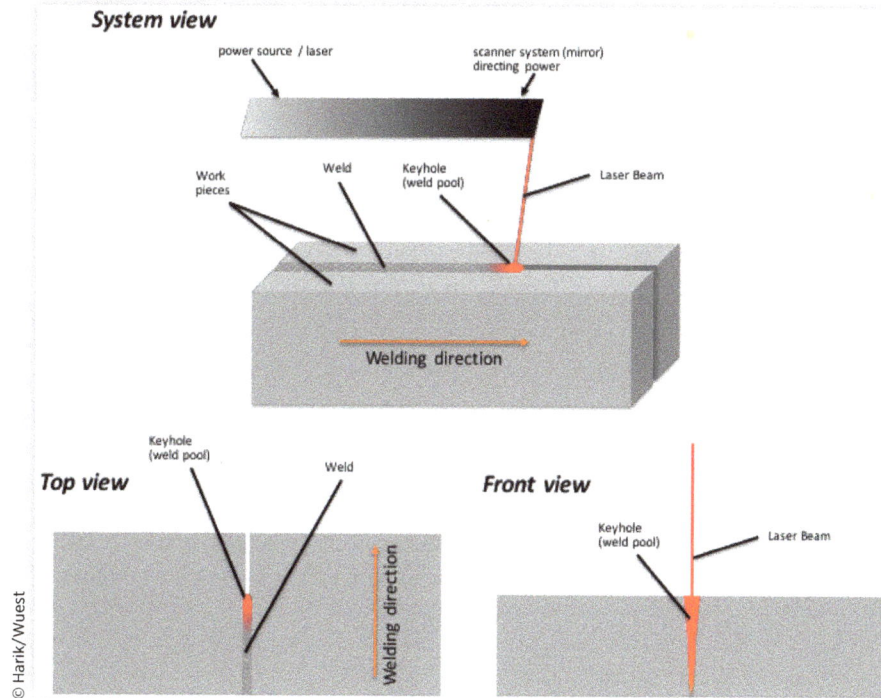

FIGURE 5.26   Schematic of heat transfer in welding process.

$$H_{welding} = V * \text{Unit Melting Energy}$$

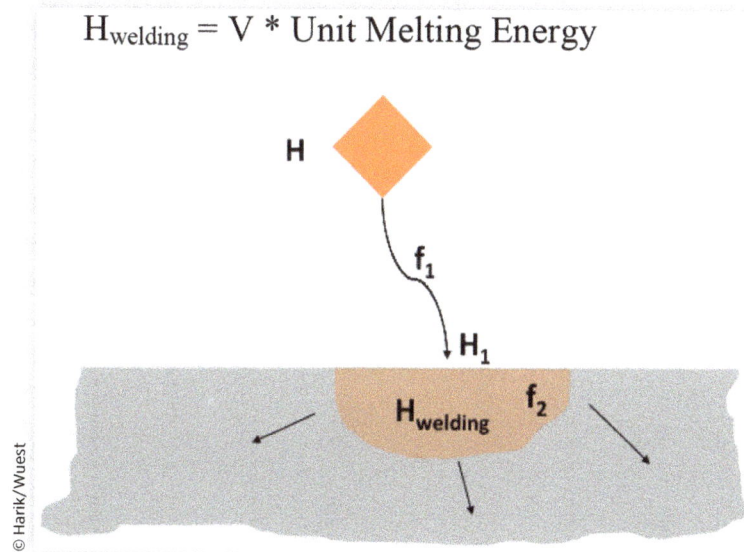

The input heat from the source (H_i), in [volt-ampere*min/mm] can be calculated based on the welding current (A) in [ampere], the arc voltage (V) in [volts], the welding speed (s) in [mm/min], and a constant. The resulting equation is

$$H_i = A * V * 0.006/s \qquad \text{Eq. (5.17)}$$

The same amount of heat is melting the welding volume V and is the function of the unit energy required for melting. The weld volume can be calculated as previously illustrated (for fillet and single V butt welds).

$$H_{welding} = V * \text{Unit Melting Energy} \qquad \text{Eq. (5.18)}$$

5.3 **Non-permanent Assembly Processes**

After the introduction to the most common permanent Assembly Processes in the previous section, we will not focus on non-permanent Assembly Processes that are widely used. Compared to the wide variety of different available techniques for permanent assembly processes, the variation in non-permanent processes seems much smaller, basically meaning fasteners and snap fits. However, when having a closer look at the variation within these overarching clusters, there is a sheer endless variation available within these clusters themselves.

5.3.1 Non-permanent Fasteners

5.3.1.1 THREADED FASTENERS

Threaded fasteners are a very common means of assembling a product. They are produced in very high numbers. We remember the *upset forging* process from Chapter 3 and thread rolling process from Chapter 2 as a common means of manufacturing threaded fasteners. However, they can also be manufactured in a variety of other ways, such as being individually turned on a lathe. Threaded fastener systems consist generally of two parts: i) the one with an external thread, e.g., screw or bolt, and ii) internal thread, e.g., nut, insert, tapped hole (Figure 5.27).

Did you know that there are left- and right-hand threads? One might ask why this is necessary as it is counterintuitive given the popular children's rhyme "Righty tighty, Lefty loosey." The reasoning behind this variation is actually manifold: it allows to i) distinguish critical connections, ii) serve at the end of a turn buckle, and iii) reduce the tendency of fasteners to get loose in certain rotating machinery, just to name a few.

Trivia: The fastener market is rather old and most see fasteners more as a commodity. However, there are interesting and very successful business model innovations that impacted this rather established market. In the following we take a quick look at the Germany-based Würth Group,* a market leader for assembly and fastener materials. Würth Group is a

FIGURE 5.27 Depiction of various threaded fasteners, e.g., bolts and screws.

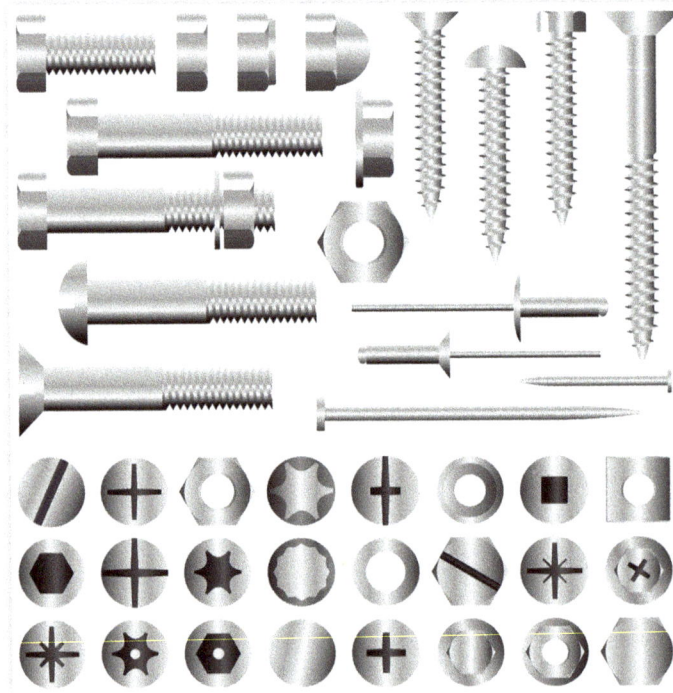

© Jason Winter/Shutterstock

---

* http://www.wuerth.com/web/en/wuerthcom/unternehmen/unternehmen_1.php.

worldwide wholesaler of fasteners, screws and screw accessories, etc., offering approximately 120,000 different products. They have over 74,000 employees (as of 2018) which makes them one of the largest non-listed company in Germany.

FIGURE 5.28 Bolts and corresponding nuts with threads.

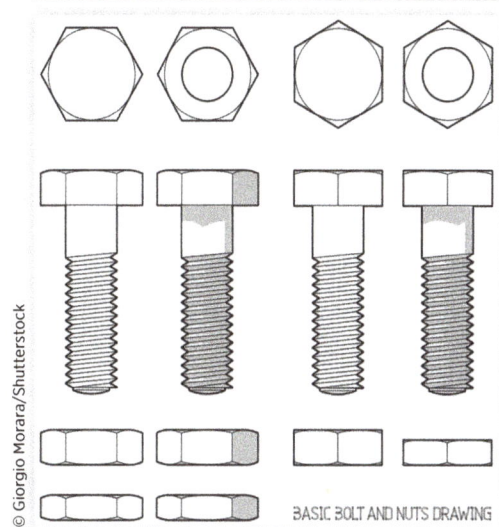

One factor that differentiates them from many competitors is their large investment in research. They filed for over 60 patents in 2007 alone. They were one of the first fastener companies that successfully adopted an eCommerce business model, increasing their effectiveness, efficiency, and customer loyalty (some call it lock-in) by increasing convenience. They are supplying customers directly at the site using and supporting state-of-the-art methods such as Just in Time/Just in Sequence (JIT/JIS) and (digital) Kanban (see Chapter 9). They supply all the equipment and storage to the customer, lowering the initial investment needed by customers to adopt their system – and they actually also offer a form of consulting service for their customers to increase their effectiveness and efficiency when it comes to fastener use. This leads to very high customer satisfaction and, as it creates true value, sustained customer loyalty. We see once more, even though a process or system, like fasteners, are available for centuries, there is always an opportunity to innovate and gain a competitive edge when thinking innovatively and using new technologies the right way.

© Giorgio Morara/Shutterstock

BASIC BOLT AND NUTS DRAWING

5.3.1.2 BOLTS AND NUTS

Bolts and nuts (see Figure 5.28) are the thread system of choice for machine builders around the world. They are available in countless materials, forms, and lengths. Bolts generally consist of a threaded area, the *thread*, and a part that allows for applying the torque to "fasten" the fastener assembly system, commonly referred to as the *head*. In most cases the head has the same external shape as the bolt's corresponding nut.

5.3.1.3 THREAD PITCH

For threaded fasteners one defining element is the thread. There are many different variations of thread styles and types available, some motivated by performance, others by historic/geographic origin. Popular and widely used thread variants include Sharp V, Metric, Square, Buttress, Whitworth standard, and many more.

The threads are classified differently depending on the terminology and classification system used. In the US customary system, the *thread pitch* is determined by the number of threads per inch. In the metric system, the pitch is provided in the thread designation and describes the distance from pitch to pitch. The metric system provides the diameter, pitch, and length of the bolt (in millimeters) in its designation. For nuts the designation defines the pitch and diameter. Let's have a look at an example: M8-1.25×45. In this case, the "M" tells us that we are looking at a metric thread designation. The "8" defines the nominal thread diameter ("Major D" in Figure 5.29) and the "1.25" describes the pitch as 1.25 mm. In metric

FIGURE 5.29 Threaded fasteners – external thread terminology (following ASME B1.7M-1984).

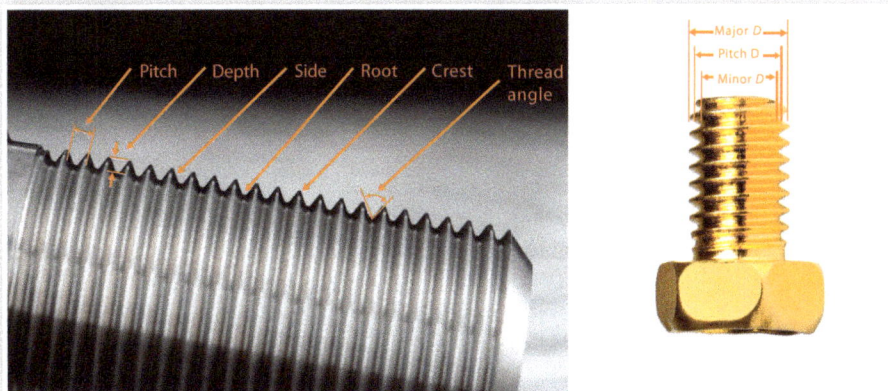

Pitch Depth Side Root Crest Thread angle

Major D
Pitch D
Minor D

FIGURE 5.30 Threaded bolt and nut.

© Oleksandr_Delyk/Shutterstock

FIGURE 5.31 Selection of different threaded metal studs.

© Prabhjit S. Kalsi/Shutterstock

bolt designations where the second number (-x.xx) is missing, the bolt has an assumed "coarse thread" which equals ~1.75mm. The "45"" describes the length of the bolt. The corresponding nut would be simply labeled M8.

The nominal diameter of the bolt ("Major D" in Figure 5.30) is generally chosen slightly smaller than the receiving major diameter of the corresponding nut to allow for assembly.

5.3.1.4 STUDS

Studs are similar to bolts, however, do not have a "head" – thus they are often referred to as "headless fasteners." They consist partly or entirely of a threaded area (see Figure 5.31). It is possible to use two nuts and in essence simulate a bolt and nut system or a tapped hole in a workpiece and a nut on the other end. However, studs can be and are used in a wide variety of applications. A very common one is as connecting rods, welded (and thus permanently fixed) to a part or as movable axis.

5.3.1.5 SCREWS

Screws are very commonly used in woodworking and applications around the house. Their main differentiation to bolts is that they do not require a corresponding nut. Like bolts, they have external threads to connect to the part. The threads for screws are often designed to be self-drilling. For metal screws, often a hole (smaller diameter than the screw) is drilled in the workpiece prior to applying the screw. However, there are several variations, for woodworking but also metals of self-drilling screws available (see Figure 5.32).

5.3.1.6 PREFERRED FAILURE MODE

We already mentioned the preferred failure mode as one reason for assemblies in general. For threaded fasteners there is also a preferred failure mode. Overall, designing the bolt based on the assembly requirement (and expected use) is essential. To successfully do so, we can not only take the strength of bolt into consideration but also have to think about the expected failure mode. Following the motto: "It's not a bug, it's a feature!"

There are three common failure modes that can occur using threaded fasteners:

- Tensile failure of bolt (*preferred* in most cases) (compare a) in Figure 5.33
- Shear failure of external threads of bolt (compare b) in Figure 5.33
- Shear failure of internal threads of nut/tapped hole

It is understood that shear failure of any kind is not desirable as it is very hard to repair a part after such a failure occurred.

FIGURE 5.32 Self-drilling screws.

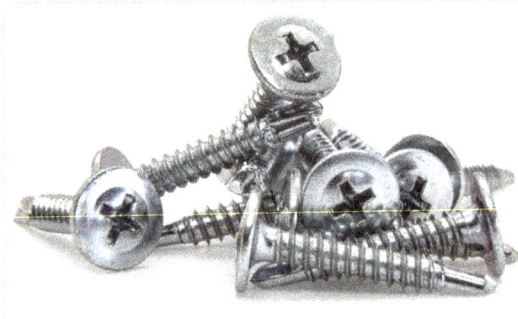

© akkara KS/Shutterstock

5.3.1.7 PINS

Pins are a means of assembly that are often used in combination with threaded fasteners. They are used to *precisely define and ensure the position* of different parts (2+) of an assembly relative to each other. There is not a defined terminology available to classify the different kinds of pins in use. However, one distinction can be made between ones that are easy to insert, sliding in the corresponding hole without much resistance, and pins that require a certain force and/or tools (e.g., hammer) to be inserted. Variants of pins are classified as locating pins, taper pins, dowel pins, center

FIGURE 5.33   Failure modes of threaded fasteners.

© safakcakir/Shutterstock; © HI JOE/Shutterstock; © Aiyoux/Shutterstock; © pattanachai w/Shutterstock

pins, grooved pins, straight pins, and many more. Just type in the search term "pins" at McMaster-Carr and browse through the variations to get an idea of the many different uses of pins. Of course there are pin classifications based on established standards like DIN or ASME available as well.

Pins can serve several different purposes in an assembly. As mentioned, they can ensure tight tolerances for precision alignment where needed. Related to that, for assemblies where precise alignment is necessary, e.g., with moving parts like a combustion engine, pins are also commonly used. Another use for pins is to provide additional support for the structure. We see that commonly for furniture assemblies (see Figure 5.34).

To manufacture pins, high precision is needed in applications where tolerances and positioning needs to be achieved. To do so, the previously introduced (Subtractive) Manufacturing processes (see Chapter 3) are used: i) for the pin we use a turning process on a lathe and ii) for the corresponding hole we employ a reaming process (for high precision/tight tolerances).

5.3.2 Other Non-permanent Assembly Processes

5.3.2.1 SNAP FIT

The last variation of non-permanent assembly processes is the *snap fit*. As the name indicates already, this process describes a part snapping on another part. The flexible part basically enables a temporary extension to "snap fit" into the final assembly. Most commonly a snap fit part is applied on plastic parts due to the required elasticity that allows for the required temporary extension. One of the most popular examples are covers for smartphones (see Figure 5.35).

FIGURE 5.34   Wooden pins used in furniture assembly.

© igorstevanovic/Shutterstock

FIGURE 5.35 Smartphone cover as typical example of non-permanent snap fit assembly.

© Marko Poplasen/Shutterstock

5.4 Recent Trends and Non-traditional Assembly Methods

We discussed several common permanent and non-permanent Assembly Processes in the previous chapters. However, as initially stated, there are many more Assembly Processes and techniques available. Some of them are more innovative with some that can be almost considered an art form. In this section, we will discuss a selection of innovative, more unusual, but nevertheless fascinating Assembly Processes.

5.4.1 Friction (Stir) Welding (FSW)

Friction (Stir) Welding is a sub-process of the Solid-State Welding Category depicted in Table 5.1. Friction (Stir) Welding is used to join workpieces that face each other by use of heat energy, generated by friction between the workpieces and the rotating tool (see Figure 5.36). The process allows to create very good joining results with high quality/high strength without distortion, harmful chemicals, or other strains to the environment. Friction (Stir) Welding is often used to join different types of (lightweight) metals or other alloys.

The tool rotates at the area where the two workpieces face each other and enters the material in the so-called entry hole. At the same time, a (large) force is applied to the tool, increasing the friction and thus the heat generated (mechanically). The high forces require heavy-duty tooling. The resulting heat energy induced joins the two workpieces without melting the materials (see Figure 5.37). It is possible to conduct Friction (Stir) Welding on a mill; however, caution has to be taken. At the end of the process, the tool is withdrawn and leaves a hole at the exit point, called the "exit hole." This "exit hole" is often seen as an additional disadvantage of the process. Today, we already see the development of new strategies that are successfully eliminating the exit hole for Friction (Stir) Welding processes (Hattingh et al., 2015), and there are also methods that fill the exit hole in a secondary process (Behmand et al., 2015).

Friction (Stir) Welding processes are mainly used in conjuncture with metal or polymer (thermoplastics) materials. The applications range from automotive to aerospace parts. Two explicit examples where Friction (Stir) Welding was used as it was able to achieve better results than other techniques, such as Spot Welding, are the Apple iMac, where joining the front and back panel precisely presented a challenge for other joining processes, as well as the Fort GT where the center tunnel was manufactured using Friction (Stir) Welding to achieve the required properties.*

---

* http://www.holroyd.com/blog/friction-stir-welding-applications/.

FIGURE 5.36   Schematic of friction (stir) welding process.

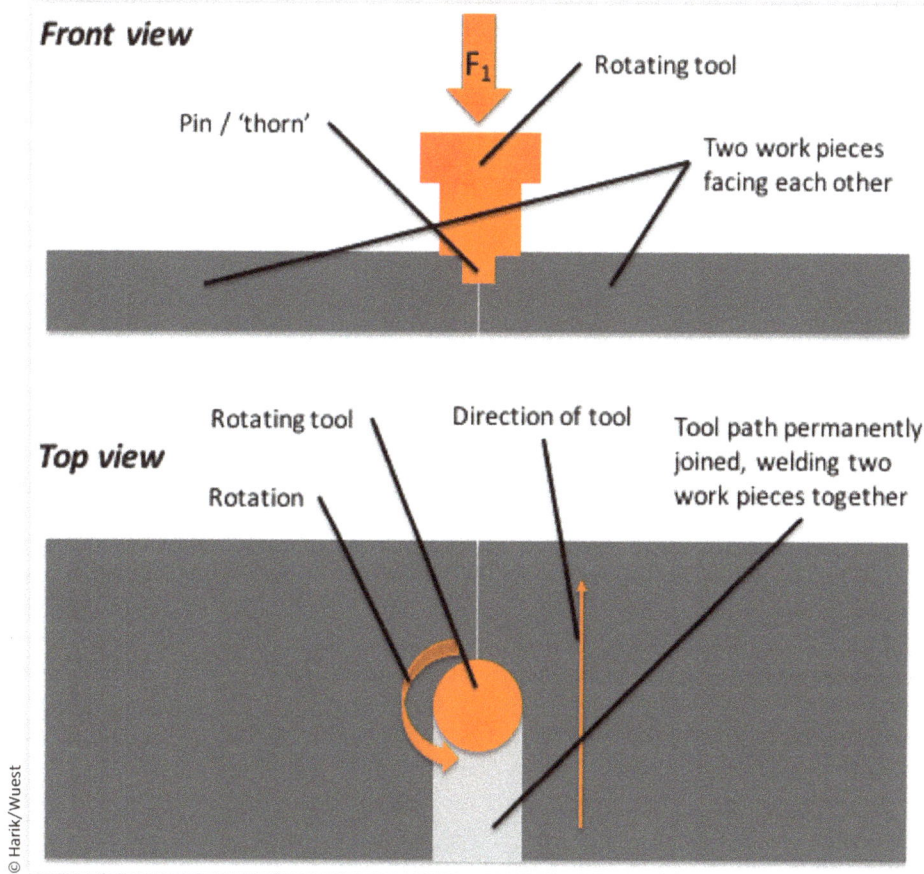

FIGURE 5.37   Industrial friction (stir) welding process.

5.4.2 Shrink/Press Fit

A *Shrink Fit* assembly describes the (in most cases) permanent joining of two workpieces through interference fit making use of the (thermal) expansion/shrinkage of (metal) materials.

Variations of Shrink Fit Assembly processes include i) heating of the receiving workpiece to expand the diameter of the insert, or ii) cooling of the to-be-inserted workpiece to reduce diameter, or iii) the heating of the receiving AND cooling of to-be-inserted part to

FIGURE 5.38 Depiction of shrink fit assembly process.

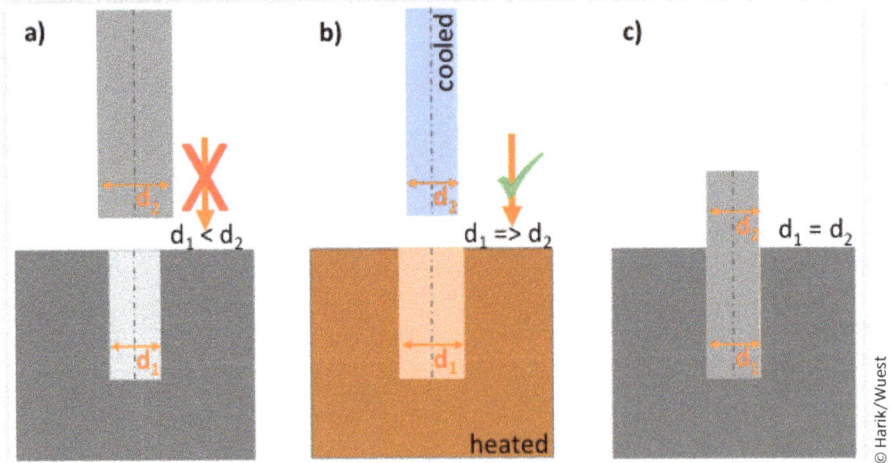

simultaneously increase the diameter of the insert and reduce the diameter for the to be inserted part (depicted in Figure 5.38b). Of the three variants, variant iii) generally achieves the tightest fit.

The three main process stages are illustrated in Figure 5.38a-c. At the starting point (a) of the process, the diameter of the receiving workpiece opening (d_1) is larger than the one of the to-be-inserted workpiece (d_2). By heating and thus expanding the receiving workpiece and cooling and thus shrinking the to-be-inserted workpiece, the two diameters allow the two parts to join ($d_1 => d_2$). While the two diameters (d_1, d_2) are becoming closer to being equal, there is often still a certain amount of force required to insert the workpiece in the receiving part.

Shrink fits are resulting in extremely high accuracy and also reduce the amount of parts required in an assembly, e.g., by reducing the need for a threaded fastener. The most common geometries used in Shrink Fit assemblies are circular shapes (hole and pin). However, there are several variations that utilize the Shrink Fit principle to join two more complex shapes through temporary (thermal) shrinkage/expansion of the parts.

5.4.3 Japanese Joinery

In woodworking, creative assembly methods and joints have been developed and used for centuries. One particularly interesting, creative, and artful variant of assembly is Japanese joinery (see Figure 5.39, Sumiyoshi and Matsui, 1989). This form of designing joints recently became popular due to its beauty and simplicity in creating secure and functional joints. In this form of joint design, the designer creatively combines parameters like i) the shape of the parts, the ii) intended function and load, and iii) (sometimes) pins to secure the assembly. Rarely additional adhesives or fasteners are used in addition.

Depending on the intended use, the joint design will focus on certain load and movements, while having degrees of freedom in other dimensions (for ease of assembly). Some of the principles used are also guiding principles in today's DfA.

5.5 Design for Assembly

High-precision assemblies need special attention during design. The tolerances (fulfilled for each individual part) may cause problems for the final assembly for systems that require extreme precision like high-performance engines, measurement systems, or precision weapons. This has to be addressed by the design engineer and special measures might be necessary aside from planning the tolerances, such as pins (see Section 5.3 Non-permanent Assembly Processes).

FIGURE 5.39 Furniture assembly using Japanese joinery.

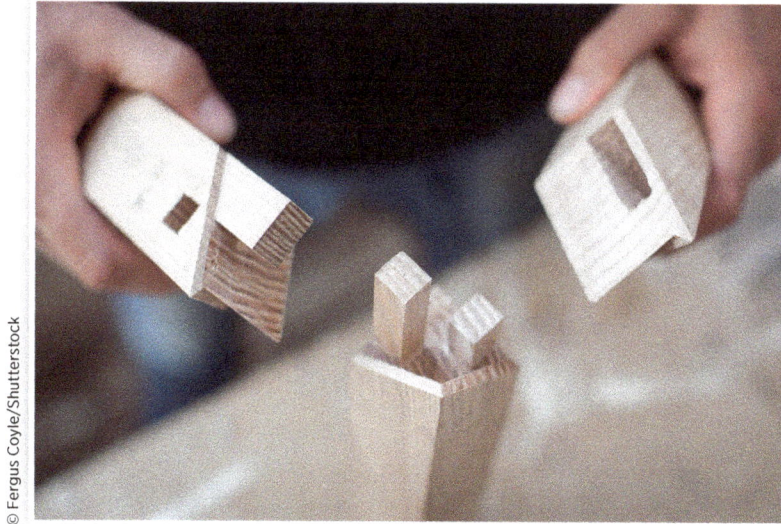

© Fergus Coyle/Shutterstock

Notes:

DfM and DfA:

- It is important to plan assembly processes during design!

- Keep assembly options in mind (both as opportunity [e.g., reduce manufacturing resource requirements] and as challenge) when planning manufacturing.

- Design "hard facts" (e.g., joint strength) but also "soft facts"(e.g., ergonomics).

References

Behmand, S.A., Mirsalehi, S.E., Omidvar, H., and Safarkhanian, M.A., "Filling Exit Holes of Friction Stir Welding Lap Joints Using Consumable Pin Tools," *Science and Technology of Welding and Joining* 20, 4 (2015): 330-336.

Dilip, J., Anam, M., Pal, D., and Strucker, B., "A Short Study on the Fabrication of Single Track Deposits in SLM and Characterization," *Solid Freeform Fabrication 2016: Proceedings of the 276th Annual International—Solid Freeform Fabrication Symposium—An Additive Manufacturing Conference*, 2016.

Everton, S.K., Hirsch, M., Stravroulakis, P., Leach, R. K. et al., "Review of In-Situ Process Monitoring and In Situ Metrology for Metal Additive Manufacturing," *Materials & Design* 95 (2016): 431-445.

Hattingh, D.G., von Wielligh, L., Thomas, W., and James, M.N., "Friction Processing as an Alternative Joining Technology for the Nuclear Industry," *Journal of the Southern African Institute of Mining and Metallurgy* 115, no. 10 (2015): 903-912.

Sumiyoshi, T. and Matsui, G., *Wood Joints in Classical Japanese Architecture* (Japan: Kajima Institute Publisher, 1989).

Wang, T., Chen, J., Gao, X., and Li, W., "Quality Monitoring for Laser Welding Based on High-Speed Photography and Support Vector Machine," *Applied Sciences* 7 (2017): 299, doi:10.3390/app7030299.

Wiegemann, M., "Adhesion in Blue Mussels (*Mytilus edulis*) and Barnacles (genus *Balanus*): Mechanisms and Technical Applications," *Aquatic Sciences* 67 (2005): 166.

Computer Aided Manufacturing

Computer Aided Manufacturing (CAM) is the art of using computer systems to create machine motion that enables the usage of equipment for manufacturing. This "tool path planning" is a necessary input for CNC (Computer Numerical Control) manufacturing equipment. It requires skill and expertise to understand the impact of decisions made during the selection of machine parameters on manufacturing quality. This chapter investigates these skills and their application in Computer Aided Manufacturing.

Computer Aided Design (CAD) and Computer Aided Manufacturing (CAM) emerged in unison with the development of numerically controlled subtractive manufacturing processes. Often referred to as the world's first CAD system, Sketchpad (Sutherland, 1964) stems from the same era as Pronto developed by Hanratty in 1957, labeled as one of the first Numerical Control (NC) manufacturing systems. These systems were both heavily influenced by existing subtractive manufacturing processes, and this influence can still be seen today. As a simple example, designing a cylinder with a CAD system testifies to this historic influence. Currently, if you want to model a cylinder using CAD tools, you can do it in one of the following two ways:

- Create a circle and extrude it along its normal; this is influenced by milling (*prismatic*).

- Create a profile and revolve it around an axis; this is influenced by turning (*revolution*).

No serious textbook with the goal of introducing readers to manufacturing can escape dealing with the topic of CAM. Most modern manufacturing equipment requires a digital input code to operate (Figure 6.1). Generating this code requires several steps including but not limited to drafting the code, simulating the code before actually putting the tool into motion, and performing a dry run to ensure that no collisions are expected

Before we delve deeper into the chapter, we will quickly touch upon when to use and when not to use CAM systems by looking into the main advantages and disadvantages of CAM systems.

CAM system has many advantages, including but not limited to process reliability, accuracy, transparency, and communication. Indeed, the ability to obtain the identical results* based on the same initial conditions is highly desirable especially in industries that require tight tolerance.

---

*"Identical results" is understood as within the acceptable range, thus fulfilling the requirements from a quality perspective. Each product/part is different/unique when we inspect it closely enough due to deviations of the process, material, or environmental factors.

FIGURE 6.1 CAD model and manufactured sheet metal.

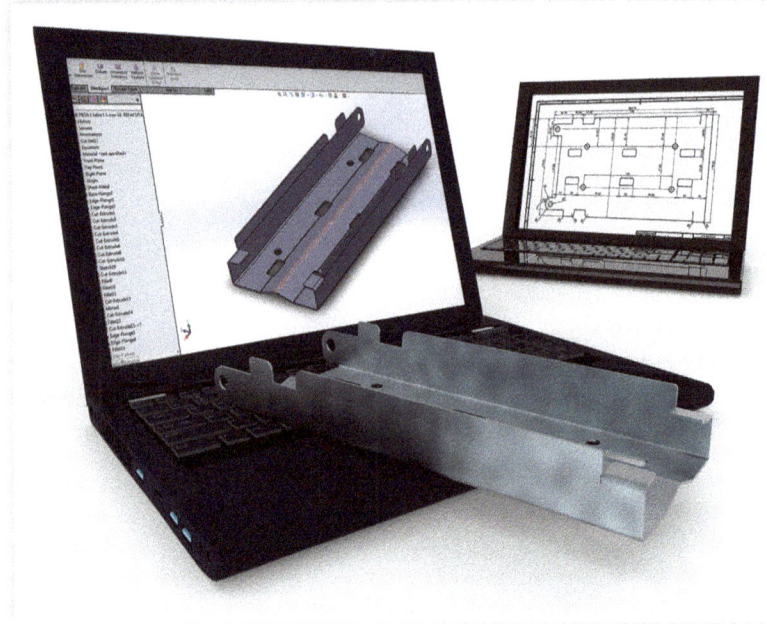

© Solcan Design/Shutterstock

This creates a process reliability that is a driving force for overall quality. This process reliability increases accuracy and helps to overcome errors found regularly in manual operations, often induced by human factors. Finally, CAM systems offer cross-cultural communication as many machine platforms follow ISO standards (G-Code) with very minor modifications depending on the controllers used.

While the advantages are numerous, leading to the adoption of CAD/CAM systems in most industries, there is still "no free lunch." CAD/CAM solutions are not perfect, and users have to cope with certain disadvantages on a regular basis. Three core disadvantages often associated with CAD/CAM systems are expertise, cost, and post-processing. Skilled and experienced programmers are hard to acquire and train. Multiple different programming platforms make the training process highly tool specific, and skills are only partly transferable across platforms. Furthermore, CAM systems are expensive and they require major investments that need to be justified. Systems easily can cost hundreds of thousands of dollars for complex engineering tasks. Finally, one of the complications is the adaptation of CAM systems to different equipment through post-processing. This "translation" step is often not trivial and requires a set of expertise on its own.

This chapter introduces the topic of CAM and provides illustrative examples of CAM applications and problems in the context of both subtractive and additive manufacturing processes. We first present the numerical chain that encompasses all the trades needed to manufacture, and their digital counterpart. Following, we detail geometric modeling, which is a precursor to appropriate and repeatable manufacturing operations. Next, we move on to present one of the most fundamental concepts in manufacturing: understanding manufacturing references and programming domains, as well as forward/inverse kinematics. We finish the chapter by introducing the standard G&M Code.

6.1 The Numerical Chain

This first section presents the numerical chain. First we focus on understanding the different eras of design and manufacturing from a historical perspective, starting from hand drawing to today's Product Lifecycle Management (PLM). Following, the two main starting points of the numerical chain are presented and each of the different trades within the numerical chain is introduced.

6.1.1 **From Hand Drawing to PLM**

This section covers the shift from hand drawings of physical products as input for manufacturing to today's sophisticated PLM context that fosters real-time collaboration across continents, enabling the design and manufacturing of highly complex systems. We can distinguish five eras toward the progress of today's tools (Figure 6.3):

- *Hand Drawing*: Initially started as an archival process for data perpetuation, hand drawing leaves room for interpretation, and concepts/designs are not set in stone. Sometimes, those who made the original drawings/sketches are the only ones who fully understand it and the implications for following processes, such as manufacturing. Da Vinci's "Automobile" is a fine example illustrating the drawbacks of hand drawings to understanding complexity (Figure 6.2). Originally designed in 1478, the design and its associated hand drawing was so complex and hard to understand that manufacturers and even researchers had great difficulty understanding the functionality and details. The "Automobile" is often considered as the first car, and the first functional prototype based on the original drawings was only manufactured in 2004. In addition to the fact that the design interpretation is somewhat difficult, changing the design required discarding the original sketch and restarting from scratch. Finally, hand drawing has very limited usefulness when it comes to three-dimensional (3D) complex shapes. It is hard to imagine how many drawings are needed to present a complex shape and ensure that everyone – and not only those who designed it – can understand and act on it to allow repeatable, high quality manufacturing.

- *2D Computerized Modeling*: When Sutherland created Sketchpad in the 1960s at Massachusetts Institute of Technology (MIT), it was mostly used to draw 2D elements such as lines and circular portions. Considered to be the father of CAD systems, Sketchpad was without a question a breakthrough in pushing the development of computer graphics to new heights. Without it, many of today's standard systems would not exist, and the way parts and products are designed might be very different. The similarity with Hand Drawing earned the software a nickname "Robot Draftsman" highlighting the automation of the draftsman's profession. Sketchpad introduced another revolutionary CAD concept that persists to this day: geometric constraints such as length and angles.

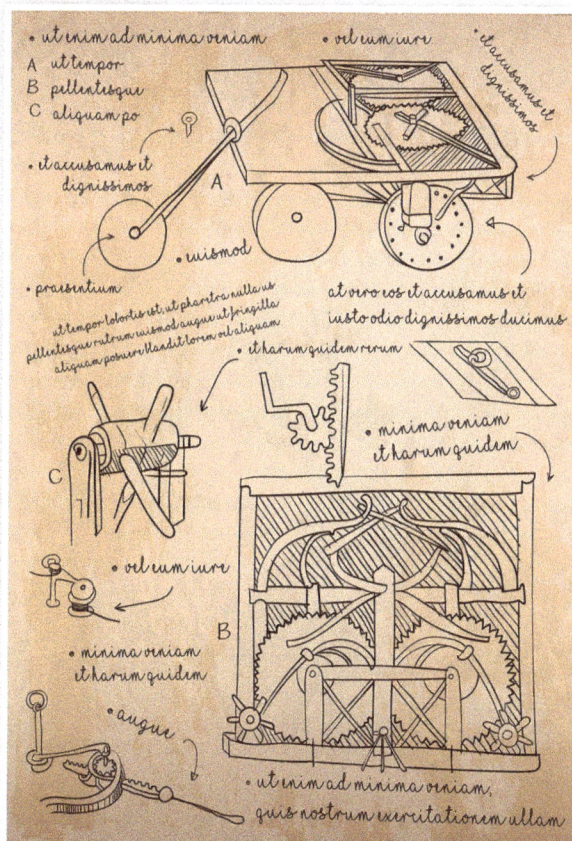

FIGURE 6.2   Da Vinci's automobile.

© Kwirry/Shutterstock

- *3D Computerized Modeling*: With the development of 3D surface representations, mostly by Pierre Bézier from France, 3D modeling became a real and exciting possibility for the digital design community. For the first time, the representation of complex geometries through mathematical formulations was possible. Section 2 of this chapter details those mathematical foundations that are the foundation of all 3D geometric representations. However, at this stage, 3D representations were still purely geometric representations with hard to manage changes, not parametric, and did not include documentation on relevant information such as specifications, materials, and such.

- *Digital Mock-Up (DMU)*: Marked with the introduction of change management and the ability to embed specifications and information within the design context. DMU is considered to have unlocked collaborative and concurrent engineering and the ability to manage functionality in a parametric fashion. This led to a massive reduction in the Design to Market (DtM) timeline from essentially years to months. This had a tremendous impact on customers' expectations and how businesses operate today. It has mostly replaced the need for a Physical Mock-up (PMU), saving significant time, resources, and cost. An example is the

FIGURE 6.3 From hand drawing to PLM.

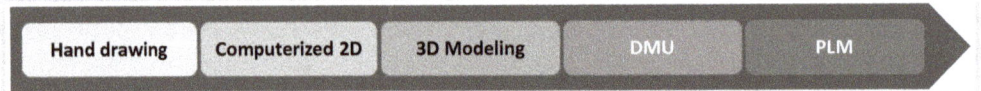

Hand drawing	Computerized 2D	3D Modeling	DMU	PLM

© Harik/Wuest

Falcon 7X business jet manufactured by Dassault Aviation (DA). Highlighting the importance of software during the design and manufacturing of the 7X, DA used the colors of the used software CATIA from Dassault Systemes (DS) in lieu of the traditional DA colors on the aircraft.

- *Product Lifecycle Management (PLM)*: The four prior eras concentrated on the so-called Beginning of Life (BOL) of a product. They represented the initial concept (ideation), specifications, design, and manufacturing of a product. Currently, we are interested in understanding the product through not only its BOL but as well its Middle of Life (MOL) and End of Life (EOL). MOL describes the usage, maintenance, and repair of a product after it exits the manufacturing industry. EOL stands for the withdrawal and remanufacturing, as well as recycling/disposal of the product at the end of its usage when it reaches the end of its useful life. While this end of useful life is often based on technical factors (technical obsolescence), there are additionally factors that initiate the transition (planned obsolescence). Data from consumer usage heavily influences the BOL, and as such, PLM offers the ability to include two dimensions in the realization of a product: across implicated trades (Design, Manufacturing, and Process Planning) and over time (BOL, MOL, and EOL).

The five historical development eras of CAD/CAM we have distinguished above are now gearing toward new concepts of enabling the product to actively sense, collect, and analyze data along the life cycle. The concepts exploring technical and economic opportunities based on this data availability are associated with terms such as "Industry 4.0," "Smart Manufacturing," and "Industrial Internet" are at the heart of current research and have not yet reached full maturity. The authors of this book believe we are on the verge of a sixth era with the introduction of smart manufacturing concepts (Chapter 10). You will all cross paths with these concepts in your careers at one point or another whether you choose to focus on design, manufacturing, or even operation of complex products.

Today's world of complex engineering problems requires this vertical and horizontal integration. Good examples are any "flying" objects! Let's have a look at a current Airbus A400M (see Figure 6.4). We can directly observe the complexity of this platform system, as the many modules and parts are designed and manufactured in a variety of different countries – many of them do not even share a common language (i.e., Spain, Portugal, UK, France, Germany, and Turkey). This additional layer of complexity emphasizes the need for a commonly accepted, standardized platform, such as our CAD/CAM and PLM systems, that allows effective and efficient collaboration across country, language, department, and business units.

Throughout this chapter, we regularly refer to CAx tools. CAx stands for Computer Aided X, where X can be substituted for D for Design (CAD), M for Manufacturing (CAM), PP for Process Planning (CAPP), and E for Engineering (CAE), among others.

FIGURE 6.4 Today's concurrent/collaborative engineering represented by the A400M.

© Shutterstock

6.1.2 **The Numerical Chain**

The numerical chain, sometimes alternatively referred to as digital chain, represents the generation of a new product across several trades, all within the context of the BOL. In principle, we are replaying the physical traditional chain by numerical computations and simulations. What used to be decision-making based on handmade physical prototypes is replaced with the digital chain and virtual immersion. An excellent example highlighting this switch is the car design industry. In its early days, several prototypes massively scaled were proposed. A few of these prototypes were selected and then made by hand at ¼ or 1/8 of the original scale. Subsequently, we move to two or three designs that gets modeled at the correct scale before making a final decision. For example, the 1:1 dimensions for the design prototypes handmade from clay, such as Eric Clapton's Ferrari. The numerical chain has replaced all or at least most of this historical process for today's products, especially for complex designs. This led to a drastic reduction in errors and time-to-market of new products and an increase in collaboration of engineers across continents.

Figure 6.5 shows the two main alternative starting points of the numerical chain: Specifications and a Physical Product. To distinguish between them, the first – *specifications* – indicate that we are starting from scratch without any prior knowledge of the to-be-manufactured product, and the second, *physical product*, is when we are recreating an existing product starting from its physical form.

Both numerical chains, starting from specifications – referred to as Numerical Chain 1 (NC1) - and initiated by an existing physical product, referred to as Numerical Chain 2 (NC2), follow the same manufacturing sequence once the model representation is successfully achieved. In fact, they differ solely in *how* the model/design is created. NC1 is initiated by documentation and specifications based on customer requirements and leads to the implementation of these specifications through CAD modeling. NC2, on the other hand, is initiated by the digitization of a physical product and then its subsequent model reconstruction. From that stage forward, both numerical chains proceed with CAPP, Prototyping, Manufacturing, and Quality Control, leading to obtaining a physical product. Figure 6.6 represents the sequence of the different trades utilized in the two variants of the numerical chain.

In the following text, we detail some of the trades highlighted above in the numerical chain flow:

- *Specifications*: The starting point of the project are the user or customer requirements which include functional behavior and expected performance indicators, among other things, such as aesthetics. Information such as the expected load, fatigue, expected usage time and intensity, and such are provided, often augmented with general information such as the goal of the project and the intended users. With the dawn of Internet of Things, there is a trend to use sensors to inform design specifications and derive requirements based on real usage information collected over the use phase of previous generations of the product. However, there are still many unknowns and barriers in place that have to be overcome to operationalize this vision. This is particularly true in the context of assembly planning and the implementation of virtual commissioning concepts. In this trade, poor specifications can lead to a nonconforming design aka "bad product quality." This directly impacts customers' satisfaction, essentially leading to unsatisfied customers and ultimately jeopardizing profits. Specifications can include rather detailed, often industry or manufacturing process specific, information that goes beyond general information items. These specific items highlight part/product positioning and mating surfaces for assembly or operational volume (i.e., volume for airway tubes in aircrafts or volume for engine compartment in a car). Materials can be suggested/provided as a constraint or can be selected during the design process. Specifications are often detailed by interdisciplinary teams to ensure that nontechnical insights are appropriately reflected.

FIGURE 6.5 Numerical chain starting point.

FIGURE 6.6 Sequence of trades within the different numerical chains.

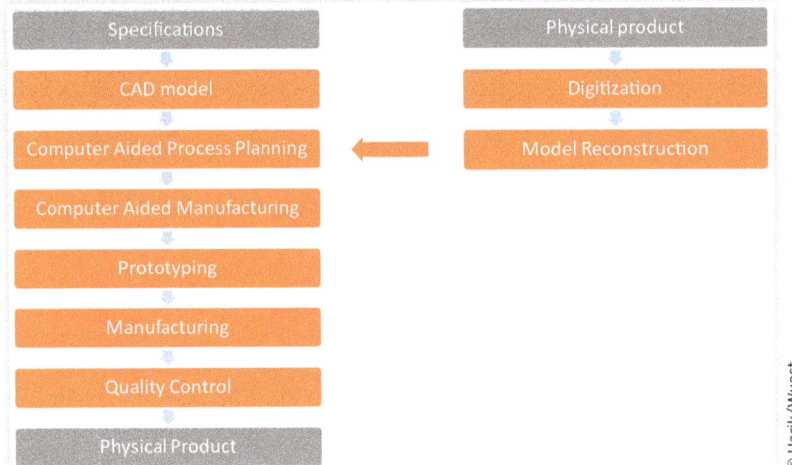

The final output of this trade is a clear document outlining required specifications as input for the design engineer.

- *Computer Aided Design (CAD)*: What used to be a mere 3D representation of the final product is now actually an involved study that interweaves cost, expected quality, sustainability, and many other parameters that influence the life of a product (hence the more recent terminology reflects this holistic perspective – PLM). Whereas design used to be performed by a handful of people, nowadays large groups of people collaborate to generate complex designs, often across borders, continents, companies, and disciplines. The CAD modeling process is often iterative and ill-defined. Furthermore, it is more often than not limited by time allocated for a certain project rather than by achieving an (close-to) optimal design. Several variables influence an appropriate design such as batch processing, resources, manufacturing constraints, and expected market price, among many others. An example of failure – in the context of achieving intended specifications – is SMART cars. Whereas the concept started by a specification where the car will be "cheap" and affordable, SMART are more perceived as luxurious cars for modern executives living in cities. Another original specification of the SMART car was scrapped before it reached production – the power train. It was originally intended as a fully electric car for use in cities. The impact of design on the final cost of a product cannot be underestimated and needs to be clearly stated: Design has one of the, if not the biggest, influences on the final cost of the product! Therefore, the time, energy, and resources invested in optimizing the design of a (reasonably complex) product is well invested and reduces headache further down the line. A straightforward example is the design of a square hole. Although achievable, having a circular hole is much cheaper to manufacture and easier to obtain as it resembles the shape of drill bits. Having a "square hole" is expensive, and unless the feature is clearly justified and has a required function, the designer should avoid this strategy. The goal of this section is to introduce design, and as such, we will limit ourselves to conclude with the DfA, DfM, and DfE concepts. DfX stands for Design for X, whereas X represents different trades. DfA stands for Design for Assembly where we take into account assemble-ability requirements. DfM stands for Design for Manufacturing where manufacturing related constrains inform the design process. DfA and DfM is often referred to in unison as DfMA. Similarly, DfE stands for Design for Ergonomics where different populations are considered and designs are adjusted to accommodate safety, comfort, etc. of the human operator or user of the product or production process. The final output of this trade is a CAD model (see Figure 6.7) of the product.

- *Digitization*: A major step in "reverse engineering" is when we "digitize" an existing physical product. This can be done for a variety of reasons. Two primary reasons for digitizing a physical product are i) creating a model for an existing physical product (to initiate NC2) or to ii) recreate a model for a manufactured product to ensure production quality. One increasingly important use case of digitization is to manufacture parts for classic cars (old timers) that are not in production and/or available any longer. By digitizing the broken part, the resulting digital model can be used as input for a modern manufacturing process such as SLS (Additive Manufacturing) to recreate a one-of-a-kind (low volume) replacement to keep the cars on the road. In both primary cases depicted before, the first step is always to obtain a cloud of points that represents the design. This generates a raw format that requires a subsequent processing step labeled model reconstruction (will be detailed next). There are a variety of mechanical and optical systems to capture model geometry available on the market with a large range in terms of ease of use, precision, and speed, just to name a few parameters. Mechanical systems, like Coordinate Measuring Machines "touch" the physical product to capture the geometry (point, line, surface). Optical systems such as 3D scanners capture the geometry from a distance without physical contact using a laser-based or any other optical or proximity system. Recent advances in the context of digitization enable even material recognition as part of the scanning process. However, this is rather novel and reliability and materials that can be recognized still needs further work to be relevant for industrial application. 3D scanning and digitization technologies can be applied at a number of different scales, ranging from submillimeter precision for individual parts all the way up to 3D scans of entire factories. Regardless of the methodology, the final output of this trade is a Cloud of Points (COP) (Figure 6.8).

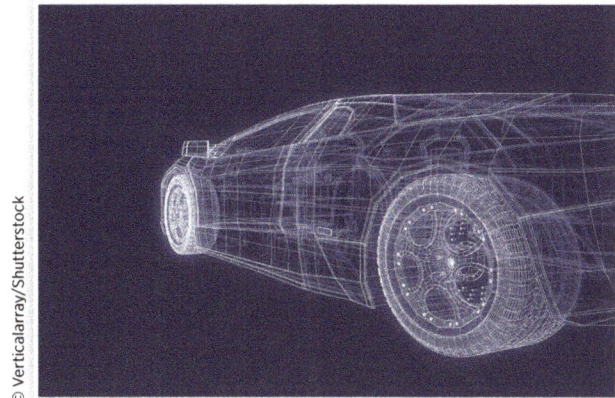

FIGURE 6.7 CAD wire frame representation.

© Verticalarray/Shutterstock

- *Model Reconstruction*: A direct and necessary step following digitization is model reconstruction. Model reconstruction recreates the physical topology and underlying geometry that represents the physical product in the virtual space. We target the fusion of the COP obtained into meaningful and representative geometrical entities/features that are required as input for our process plan/manufacturing later on. Multiple automated referencing systems support the unification of COP. Typically, an initial treatment is needed to remove different kinds of noise and undesired scans (sometimes referred to as ghosts) acquired by the system and represented in the raw COP data. Recently, PLM tools have advanced surface recognition algorithms that are automatically performing some of the traditionally manual COP treatment steps. The final output of this trade is a CAD model.

- *Computer Aided Process Planning (CAPP)*: Matchmaking design attributes with suitable manufacturing processes and technologies is the main purpose of Process Planning. Process Planning is one of the most critical jobs within the numerical chain and in the end within every manufacturing company. CAPP is essentially where both numerical chains "meet" to continue towards manufacturing and release of the final physical products. Process Planning requires human expertise and analysis of the part to be manufactured which often proves to be complex and time consuming. CAPP aims to reduce the amount of human intervention required; however humans are still indispensable in planning the manufacturing process. The part itself is analyzed in the context of available (manufacturing and other) resources. Neglecting machining difficulties, or not identifying critical ones, has detrimental consequences on meeting the deliverables specified by the part designers based on customer specifications. Therefore, problems during Process Planning increases the risk of producing products that do not meet the quality requirement. In addition, to ensure producability and quality, CAPP often involves various simulation tools that allow users to experiment with various manufacturing strategies before deciding on an optimal approach. CAPP itself varies significantly depending on what manufacturing process is used or envisioned. There are different CAPP functions and strategies in Subtractive, Deformative, and Additive Manufacturing. For Subtractive Manufacturing, functions such as optimal manufacturing mode (flank/end), tool selection, and machine accessibility are primary and govern the downstream of the numerical chain. For Additive Manufacturing, bridging, nesting, and optimal build orientation are on top of the requirements priority list. The final output of this trade is a detailed Process Plan that has a definitive list of resources needed to perform the manufacturing steps.

FIGURE 6.8 Scanning and digitization of a physical product.

© Monkey Business Images/Shutterstock

- *Computer Aided Manufacturing (CAM)*: CAM is the translation of the process plan and process parameters into toolpath trajectories. Several industries embed CAPP as a portion of CAM, to demonstrate their dual dependability. Within CAM, the target is to select the optimal tool path that minimizes both manufacturing defects as well as production time. This often requires post-processing to translate a standard obtained APT (Automated Programmable Tool) to NC commands. The NC Code is typically represented as G-Code which is quasi-industry standard. There are few examples of industries and controls that integrate a modified version of G-Code (minimal modifications). However, even in those cases, the main functions remain somewhat generic across all applications. The verification of machine simulation, accessibility, and collision avoidance is included within most CAM packages. The final output of this trade is a functional NC Code that can be sent to the CNC machine tool to initiate the manufacturing process. The NC Code represents the toolpath trajectory to manipulate the physical part depending on the chosen manufacturing process.

- *Prototyping*: This intermediate step describes the initial, small-scale, often one-of-a-kind manufacturing of the designed shape for testing purposes. The objective of prototype testing can be product focused, but can also provide insights on the manufacturing or logistics processes. The goal of prototyping is to have a "first feel" of the part, without paying too much or spending a lot of time on manufacturing. We often use low-cost materials and accept reduced tolerances to verify the general functionality of the product, e.g., drag coefficient in a wind tunnel, or the customer experience, e.g., ergonomics of the cockpit. This is a very important step before confirming manufacturing plans and is frequently considered as a mandatory "caution" milestone that can actually send design revisions all the way back to the model creation. Advancements in 3D printing and Additive Manufacturing have made prototyping a much faster process and as such have become the preferred prototyping methodology, hence the past term "rapid prototyping" that was used for certain AM techniques (Figure 6.9). Nevertheless, CNC routers for milling and turning (Desktop CNCs) are very common and widely adopted for prototyping purposes. The final output of this trade is a low-cost replicate of the final physical product. There might be different prototypes for different testing purposes, such as a design prototype – focused on the detailed outer shape – or a functional prototype, replicating the future functionality with simplified designs, etc.

- *Manufacturing*: Using the appropriate manufacturing resources, we implement the approved process plan to obtain the final physical product(s). Manufacturing has two dimensions: Physical Manufacturing and Manufacturing Planning. The first dimension is the actual physical process that leads to the manufacture of one component/a batch of components. Manufacturing Planning (the core trade of Industrial Engineers) is the dimension where we study production systems and detail the requirements planning. Examples of production systems are push, pull, and kanban, centrally controlled through MES or ERP systems, and many more. In the context of this book, we are focusing mainly on the first manufacturing dimension – Physical Manufacturing. The second dimension – Manufacturing Planning – can be better understood through selected sections of Chapter 9, Manufacturing Quality Control and Productivity. The final output of this trade is a manufactured physical product defined by the specified tolerances and correct material(s).

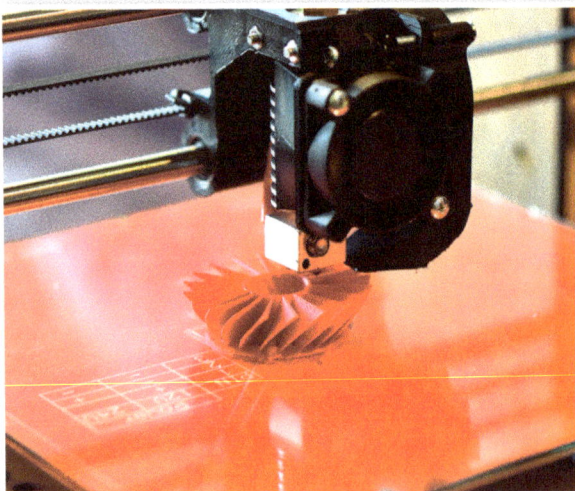

FIGURE 6.9 Rapid and low-cost 3D printing using FFF for prototyping.

© MarinaGrigorivna/Shutterstock

- *Quality Control*: The final step of the numerical chain is to implement an inspection process to ensure quality of the manufactured goods. This can be as simple as a visual inspection of the final part all the way to the more complicated methods such as the Wing Flex (where the aircrafts wings are flexed until they break!). We often resort to sampling methods in most cases of mass production, and we use individual control of each part with in situ/ex situ and/or destructive/non-destructive testing. In situ describes an inspection process that does not interrupt the process, e.g., the part is inspected while moving along on the assembly line. Ex situ on the other hand, takes the product out of the process, e.g., to an inspection station and returns it in the process after inspection is finalized. Non-destructive testing does exactly what the name indicates, it allows inspection of a part without altering them physically (aka maintaining the integrity of the part), such as the visual inspection mentioned before. Destructive testing sacrifices/destroys the product (aka the product cannot be sold after inspection) to ensure it meets the quality standards. Common sense dictates that in situ, non-destructive testing is preferable as no scrap is produced and the manufacturing process is not interrupted. However, there are quality aspects that cannot be tested in situ or non-destructive, such as residual stress allocation within a metal part. Statistical quality tools (presented in Chapter 9) such as control charts and 80-20 histograms to classify origins of defects and inaccuracies are employed in quality control divisions. More elaborate investigations include design of experiments. Innovative manufacturing environments adopt zero defect policies. The final output of this trade is a quality control check for the manufacturing part.

6.2 Geometrical Modeling

This section illustrates the fundamentals of geometric modeling to represent curves and surfaces that can be manufactured. Traditionally used as a method to "see" the components, we connect the geometric modeling to the ability of evaluating and updating designs to ensure manufacturability in this section.

Geometric representations can be explicit, implicit, or parametric. Both explicit (y=f(x)) and implicit (f(x,y)=0) can lead to singularity points and have the curve reconnect itself, and as such they are not suitable for manufacturing. Parametric (x=f(t);y=f(t)) is the most suitable geometric representation for manufacturing.

In the following sub-sections, we first present the concepts of topology and geometry before we highlight graphical representations next. Then, we touch upon several sections of geometric modeling through interpolation, approximate, and freeform (Figure 6.10).

6.2.1 Topology vs. Geometry

All solid models and 3D representations have both geometrical and topological components, in addition to parametric and specification data. The separation between "Geometry" and "Topology" is important, as geometry represents elements that are "endless" without boundaries, whereas topology represents the connectivity between multiple entities within the model. Geometry is what defines the underlying mathematical equations that generate the form.

To better understand the concepts of topology and geometry, which both are heavily correlated with their dimensions (3D, 2D, 1D, 0D), we detail Figure 6.11:

- Geometry has Surfaces (2D), Curves (1D), and Points (0D).

- Topology has Faces (2D), Edges (1D), and Vertices (0D).

FIGURE 6.10 Structure of modeling techniques.

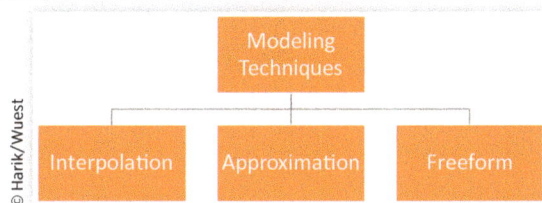

FIGURE 6.11 Topology vs. geometry.

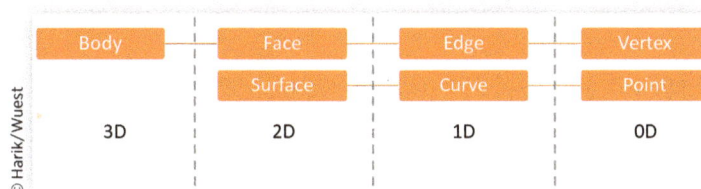

FIGURE 6.12 CSG example.

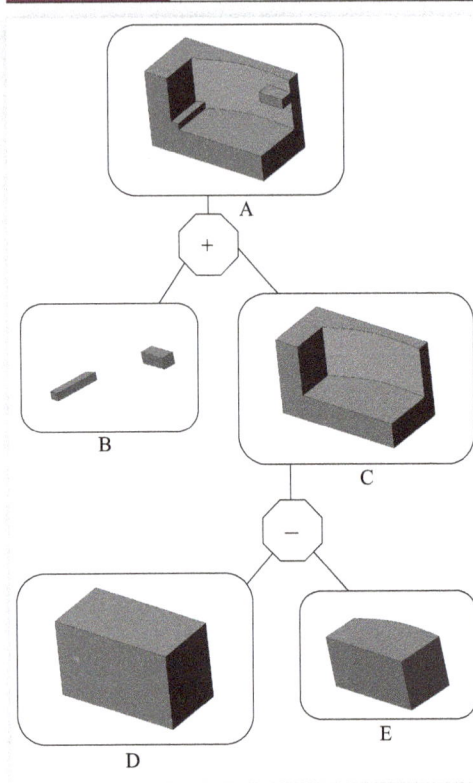

A body (or any 3D shape) is a topological element that is finite, limited, and well defined. Recently, we refer to 4D shapes as the transition of a 3D shape through the dimension of time.

An accurate definition, connecting topology and geometry states: *A topological element of dimension N is a geometrical element of dimension N bound by topological elements of dimension N-1*. For context, we provide the following example: A face is a surface bound by edges, an edge is a curve bounded by vertices.

6.2.2 Graphical Representations

To represent geometrical elements, several graphical representations and modeling techniques are found in the current literature. Representing all variations of graphical representations in detail would require a book of its own, and as such we limit ourselves to the two most prominent graphical representations of volumetric elements in CAD system: Constructive Solid Geometry (CSG) and Boundary Representation (B-Rep).

Constructive Solid Geometry (CSG) is when geometry is created through the use of Boolean operations on primitive geometrical shapes. We apply Boolean operations such as Intersection, Union, Complementation, Exclusive Union, and multiple others. The model illustrated in Figure 6.12 is an example of CSG where we subtract Element E from Element D to obtain Element C. Similarly, we create a union between Element B and Element C to obtain the final model A. The model stores both final topology and geometry and typically presents itself in a design tree.

Boundary Representation (B-Rep) is the most common methodology for representation of 3D objects and product modelling. Models consists of faces, edges, vertices, loops, and handles. Each one is defined according to the topology definition presented in the previous section. While faces, edges, and vertices are the standard topological components, a loop is a hole in a face, and a handle is a through-hole in a solid body (Figure 6.13).

To validate B-Rep models, Euler-Poincaré characteristic can be used to ensure that there are no self-intersecting elements. The governing equation is

$$F - E + V - L = 2(B - G) \qquad \text{Eq. (6.1)}$$

where
  F is the number of faces
  E is the number of edges
  V is the number of vertices
  L is the number of inner loops
  B is the number of bodies
  G is the number of genus

FIGURE 6.13 Boundary representation.

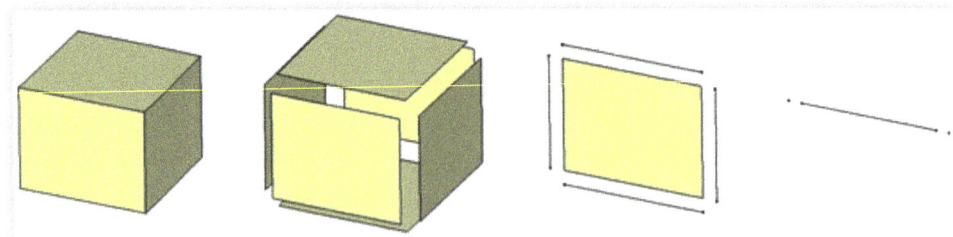

© Harik/Wuest

Typically, there are no inner loops and we have one Body and no Genus, as such a simplified formulation can be

$$F - E + V = 2 \qquad \text{Eq. (6.2)}$$

Looking at the cube above, we have 6 faces, 12 edges, and 8 vertices: $F - E + V = 6 - 12 + 8 = 2$.

6.2.3 Modeling: Interpolation

In the first of three sections in this chapter detailing geometrical modeling concepts, we briefly introduced interpolation techniques. Interpolation modeling gives a means of obtaining a curve that passes exactly through all of the given points. This can often result in a mathematical representation that passes exactly through the given points, but with heavy oscillation. As such, interpolation representations are often not smooth and, as such, are not at all desirable for manufacturing, especially Subtractive Manufacturing where oscillation leads to serious quality issues if not addressed probably in the machine control. The most prominent methods for interpolation are Lagrange and Spline. Lagrange interpolation generates a polynomial of a high order equal to the number of points used in the interpolation (minus 1). It is desired to use Lagrange for interpolations of less than 20 points. The Spline method, originating from naval construction techniques, is more accurate than Lagrange since it uses several polynomials to scan through the points. This is often labeled as piecewise polynomials. As such, Spline can handle interpolations of more than 20 points. In general, both interpolation techniques are considered to be "rigid" where curve smoothness is neglected on the expense of accuracy.

$$P(x) = \sum_{i=0}^{n} L_i Y_i \qquad \text{Eq. (6.3)}$$

where
  L_i is the Lagrange polynomial at point i
  Y_i is the value of the coordinate at point i

The Lagrange polynomial is computed as

$$L_i(x) = \frac{\prod_{j=0;j\neq i}^{n}(x - x_j)}{\prod_{j=0;j\neq i}^{n}(x_i - x_j)} \qquad \text{Eq. (6.4)}$$

6.2.4 Modeling: Approximation

In approximation techniques, we sacrifice accuracy for curve smoothness and manufacturability. The representative polynomial passes near the given points and designers are required to select the approximation degree of the geometrical model. One of the most common methods of approximation techniques is the Least-Squares method that is centered on error minimization between the proposed curve polynomial and the given set of points (Figure 6.14).

FIGURE 6.14   An approximation curve with starting points and errors (gray).

© Hanik/Wuest

To approximate a curve using one polynomial, least squares has five steps: (1) selection of polynomial degree, (2) error computation at each point, (3) computing the error function (sum of squared errors), (4) differentiation of the error function with respect to each unknown, and (5) resolution of the system of equations.

6.2.5 Modeling: Freeform

Freeform modeling is today's de facto standard for CAD/CAM systems and the most widely used representation for complex surfaces. Pierre Bézier is considered to be the originator of Freeform modeling with the famous Bézier Curves that are considered a precursor for later parametric modeling. The concept behind Bézier Curves is that points are understood as "control" elements that influence the shape/propagation and do not belong to the actual model (Figure 6.15). The mathematical formulation by Bézier has several desirable properties such as invariance to rotation, translation, and mirroring, as well as the ability to perform local modifications to influence local manufacturing setups. Freeform modeling is generally considered the most suitable technique for both design and manufacturing.

Following, we present briefly how to compute a curve using Bernstein's formulation of Bézier Curves. The computation of a parametric curve (both Y and X function of u, and not Y function of X), requires three steps: (1) Computation of Parameter Matrix, (2) Computation of Bernstein Matrix, and (3) Computation of Points Matrix. The multiplication of the three aforementioned matrixes results in the curve or surface modeling.

For the computation of the Parameter Matrix, we first count the number of controlling points that defines the contouring polygon. The number of controlling polygons' sides **n** defines the degree of the final polymer. The Matrix has the following linear form:

$$U = \begin{bmatrix} 1 & \ldots & u^i & \ldots & u^n \end{bmatrix} \qquad \text{Eq. (6.5)}$$

For the computation of the Bernstein matrix, we use the following formulation:

$$B_i^n(t) = \frac{n!}{i!(n-i)!} t^i (1-t)^{n-i} \qquad \text{Eq. (6.6)}$$

Below are the Bernstein matrices for B_1, B_2, and B_3.

$$B_1 = \begin{bmatrix} 1 & 0 \\ -1 & 1 \end{bmatrix}; \quad B_2 = \begin{bmatrix} 1 & 0 & 0 \\ -2 & 2 & 0 \\ 1 & -2 & 1 \end{bmatrix}; \quad B_3 = \begin{bmatrix} 1 & 0 & 0 & 0 \\ -3 & 3 & 0 & 0 \\ 3 & -6 & 3 & 0 \\ -1 & 3 & -3 & 1 \end{bmatrix} \qquad \text{Eq. (6.7)}$$

FIGURE 6.15 Freeform curves (solid black) are a direct result from the control points (orange) and controlling polygon (gray).

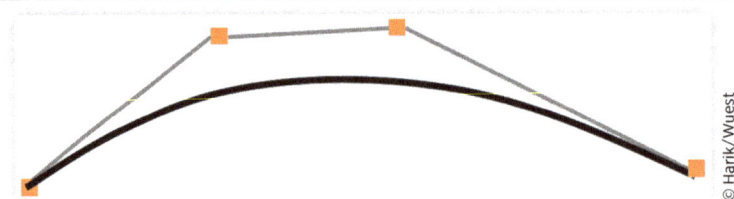

© Harik/Wuest

Finally, the control points matrix S can be filled as follows:

$$S = \begin{bmatrix} x_1 & y_1 \\ \cdots & \cdots \\ x_i & y_i \\ \cdots & \cdots \\ x_n & y_n \end{bmatrix}$$

Eq. (6.8)

Finally, the multiplication of the three above matrixes U, B, and S generates the final equation of the curve polynomial P. For more details, Homework 3 presents detailed calculation of a freeform curve and certain parameters.

6.3 Manufacturing References

Manufacturing references and their understanding is the root cause of the most expensive mistakes in manufacturing. We often model parts in a certain reference system (A), and we manufacture in a different reference system (B). There needs to be correlation and transformation between the different coordinate systems in order to ensure quality and safety, while accounting for several other parameters to ensure the part and the machine will not enter into a damaging collision. Failure to align the different reference systems and resulting collisions can result in expensive maintenance and repair costs. Moreover, this can put production schedules down and lead to significant delivery delays.

The coordination of multiple manufacturing references requires a fundamental understanding of fixed vs. variable references. The basic concepts are the same for all kinds of manufacturing; however we illustrate manufacturing references below within the context of turning machines (lathes).

Tool Origin and Chuck Origin (highlighted in orange in Figure 6.16) are fixed references that do no change when we load a new part and/or new tools. They are set by the manufacturer and are neutral positions. In contrast Tool Correction and Programming Origin need to be set with every part and/or tool change. Each time we load a new tool in the magazine holder, we need to correct the tool origin and "inform" the machine of where actually the tool tip point is located (Tool Correction). This ensures that we are performing the turning operation with the tool tip and not with the tool holder. Tool correction, although only added at times of changes and modifications with respect to new tools, need to be verified at several stages to account for tool wear.

FIGURE 6.16 Referencing in lathe machining.

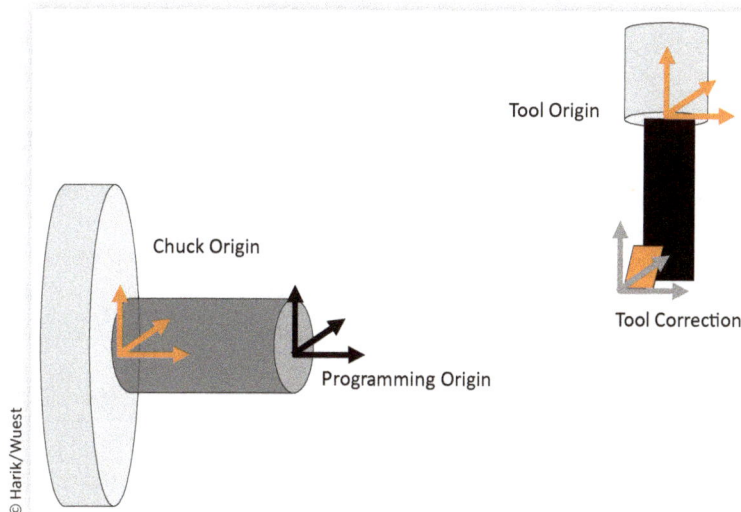

© Harik/Wuest

The last, and most important, step in geometric transformation is the programming coordinate system. When we place a new part in a machine, we program the manufacturing operation (aka tool path) with respect to a reference that is virtual. We need to inform the machine of this Programming Origin and programming reference. This step is needed every time we load a new part in any machine.

6.4 G-Code

This final section of our CAM presentation forwards a presentation for G-Codes, sometimes referred to as G&M Codes. Everything we compute to develop the tool path is transmitted through a list of standardized, coded instructions that the equipment translates using NC units to machine movements and actions. G-Code movement commands send defined instructions such as linear or circular motions, as well as possible embedded more complex manufacturing cycles, such as drilling or tapping. Not all G-Code commands are movement based, some commands perform non-movement actions such as calling a certain tool, turning on machine lubrication, or setting the unit system (Metric or Imperial) (Figure 6.17).

6.4.1 G-Code Structure

G-Code programs are structured into three main components: (1) Initialization, (2) Motion/Action, and (3) Ending. Initialization commands include lines that decide on code units, set the machine general parameters, define appropriate coordinate system, and call appropriate tools. Motion/Action commands include generating the manufacturing operation as well as deciding on cutting/adding conditions. Finally, the Ending sequence is responsible to ensure the retreat of the tool holder (or end effector in case of Additive Manufacturing), shutting down motion and speeds.

It is appropriate to state that the above terminology is more accurate in the context of Subtractive and Deformative Manufacturing. The term "tool" refers to the "die" in the context of Additive Manufacturing. The tool is what end effectors deposit materials on. Readers can refer to Chapters 3 and 8 for more information on these specific concepts and terminology.

FIGURE 6.17 G-Code was initially developed at MIT Servomechanisms Laboratory in the 1950s.

6.4.2 **G-Code Terminology**

The below listing defines the different items encountered in G-Codes:

- N: Block Number
- G: Preparatory Function
- X: Movement along the X orientation
- Y: Movement along the Y orientation
- Z: Movement along the Z orientation
- F: Feed rate
- E: Extrusion Rate
- M: Miscellaneous Functions
- S: Spindle Speed
- T: Tool Management

While most of the codes above are often followed by a number that represents a value, i.e., F1000 for a feed rate of 1000 mm/min (in case metric units are set), G&M Codes are actually followed by a number that represents an action, i.e., M6 is for tool change, or M30 for program stop. We list below the main geometrical G-Codes:

- G0: Rapid Linear Motion (does not require setting a federate)
- G1: Work Linear Motion (require setting a federate)
- G2: Clockwise Circular Motion
- G3: Counterclockwise Circular Motion

6.4.3 **G-Code Example**

To better illustrate and understand G-Codes, we present the following code. Details of the code are presented line by line, within the code comments (after the semicolon[;]).

%

O01114; This represents the Code Name that you will call and activate

G20 G21G40 G80 G90 G58; G20 Switch to imperial units (inch) in contrast to G21 which is metric units (mm), G40 Turn cutter compensation off, G80 Cancel canned cycle, G90 Use absolute coordinate system (in contrast to G91), G58 Work coordinate system that needs to be defined based on the part

G00 G90 G53 Z0; G00 Rapid linear move, G90 Use absolute coordinate positioning, G53 Reference coordinate system from machine home, Z0 Move to z=0

(OPERATION SURFACING)

M06 T4; Change to Tool 4 or Call Tool 4

G43 H04 D04; G43 Create offsets from the positive direction, H04 Length by the value calibrated in H04, D04 diameter by the value calibrated in D04

G00 G90 G58 G00; Move rapidly, G90 to absolute coordinate system, G58 work coordinate system previously defined based on the part

G00 X-0.5 Y-2. S1500 M03; Rapidly move to x=-0.5 and Y=-2 at a surface speed of 1500, M03 Start the spindle in the clockwise direction M3, in contrast to M4 which is the CCW

G00 Z3.; Rapidly go to Z=3

M08; Turn on coolant (flood setting), in contrast to M07 which is the mist setting

G01 Z-0.05 F30.; G01 Z-0.05 Move linearly to z=-0.05, F30 At a feed rate of 30

G01 Y2.; Move linearly to Y=2 still at feed rate of 30

G01 X0.5; Move linearly to x=0.5

G01 Y-2.; Move linearly to y=-2

M05; Stop the Spindle

M09; Turn off the coolant

G00 G90 G53 Z0 G90; Use absolute coordinate system, G53 Reference coordinate system from machine home, Z0 Rapidly go to z=0

M01; Program Stop

(OPERATION DRILLING)

M06 T3; Change to Tool 03

G43 H03 D03; Call offsets of Tool 3 and Tool compensation

G00 G90 G58; Ensure we are operating in the G58 Work Coordinate System in absolute

G00 X0 Y1. S5000 M03; Rapidly go to x=0, y=2 with a surface speed of 5000, M03 Start the spindle in the clockwise direction

G00 Z3; Rapidly move to z=3

M08; Turn on coolant

G98 G81 Z-1.25 R0.1 F20.; G98 Retract tool to starting z height, G81 Start basic drilling cycle where tool drills in at feed rate and removes itself rapidly out, Z-1.25 drill in to z= -1.25, R0.1 With drill retract.1, F20 at a feed rate of 20

Y2.; Go to y=2

G80; Cancel canned cycle

(END CYCLE)

M05; Stop the Spindle

M09; Stop Lubrication and turn coolant off

G00 G90 G53 Z0 G90; Use absolute coordinate positioning, G53 Reference coordinate system from machine home, Z0 Go rapidly to z=0

G00 G90 G53 Y0 G90; Use absolute coordinate positioning, G53 Reference coordinate system from machine home.Y0 Go rapidly to y=0

M30; End Program
%

Reference

Sutherland, I.E., Sketchpad: A Man-Machine Graphical Communication System,' Ph.D. thesis, MIT, 1964, https://www.cl.cam.ac.uk/techreports/UCAM-CL-TR-574.pdf.

7

Polymers Manufacturing

This chapter collates manufacturing techniques that are relevant to polymers. Some of these techniques can be considered Deformative Manufacturing (Chapter 2), some Additive Manufacturing (Chapter 4), and also Subtractive Manufacturing (Chapter 3) are core processes that are regularly used in conjuncture with manufacturing polymer parts. Of course, polymer parts are also used in Assembly Processes (Chapter 5) and planned using CAD/CAM (Chapter 6). In this chapter, we highlight the specifications, spin, and nuances that polymers as the input raw material put on manufacturing processes. You may know polymers under the commonly used names plastics, rubbers, or nylons. We will discuss the different terms and what they imply later in this chapter.

Almost every item that is in use contains polymers: from household items (smartphones, televisions, furniture, kitchen appliances) to transport elements (aircrafts, cars, ships), as well as medical devices and literally items from any other industry known to mankind. Your wallet in your pocket contains polymers (or credit cards), electric wires are insulated with polymers, and your favorite childhood toy (often Lego for engineers) is made of polymers. Their wide application, horizontal, and vertical distribution make the polymers very important for the national and global economy.

The ease of recycling and shaping as well as their comparably low cost, material properties, and availability helped propel polymers to gain the lion's share of all manufactured goods. Earlier polymers relied on organic materials, and only a little over 100 years ago, in 1907, the process to produce polymers artificially was accidentally discovered. The meteoric rise of plastics and its omnipresence in our daily lives however started when the industrial process of creating polymers from oil was discovered and applied on an industrial scale. Since 2010, according to americanchemistry.com, the plastics industry has invested nearly $47 billion in the United States. Over 460 plastics processing projects have been since announced in the USA, and the net plastic exports are expected to grow from $6.5 billion in 2014 to $21.5 billion in 2030. All of these competitive advantages led to a 20% industry employment stemming from plastics industries. Overall, a total of 2.7 million jobs are expected to be added that are supported by the US plastics industry. Outside of the United States, both China and Mexico are the leading export destinations according to the American Chemistry Council.

Polymers are nowadays often perceived as a strain to the environment with polymer parts polluting our oceans and beaches and micro-plastics found in fishes and unfortunately have been found in humans as well. Several policies have been put into action that target the reduction of the use of plastics, especially single-use polymers. Media attention often hovers around plastic straws, plastic bags, plastic bottles, and expanded polystyrene cups. Expanded polystyrene cups are often referred to as Styrofoam cups; however, "Styrofoam" is actually a trade name for building materials owned by Dow. Several countries and states have banned plastic bags already, and the European Union (EU) just recently announced an EU wide ban of single-use plastics, expected to be in effect by 2021. This might further increase the market share of recyclable thermosets compared to the single-use thermosets and elastomers.

To illustrate the rapidly growing usage of polymers in recent years, we will use an example from the automotive industry. Before the 1960s, the usage of polymers in the automotive industry was little to none existent. In contrast, polymers make up one of the most used engineering materials in today's cars: rear body panels, roof rails, wipers, headlight casing, grille, front bumpers, tires, mud guards, rear bumper, switches, and countless other parts are made of polymers (Figure 7.1). While a major factor for this increase is grounded in the lower cost, lower weight, and ability to mass produce complex-shaped polymer parts, there are also other considerations supporting this growth. One example are bumpers, traditionally made of steel or other (sheet) metals, they had the disadvantage of once being deformed, they stayed deformed as well as being considered a hazard for pedestrian safety. Modern polymer bumpers make use of the elastic deformation capability of selected polymers. The Smart car, for example, claims that the bumper will not be damaged (structurally at least – scratches in the paint might occur) when "bumping" in another car with a speed of less than 5 km/h.

Polymers Manufacturing techniques often include three distinct stages: (1) Mixing of Materials, (2) Forming of Polymers, and (3) Finishing. For *Stage 1*, we typically start with polymers in powder or pellet form that are then melted and rendered in a viscous state to facilitate flow and forming. It is at this stage where we typically insert additives such as toughening agents and colorants. *Stage 2* is where the actual forming of the viscous material takes place through either two-dimensional (2D) manufacturing (extrusion) or three-dimensional (3D) manufacturing (injection and blow molding). Stage 2 can be comprised of several steps and is not limited to one single forming technique. Finally, *Stage 3* includes post-processing such as machining, drilling, and finishing of the manufactured product, as well as the assembly of multiple components into the final product.

This chapter starts with introducing polymers themselves, presenting the definition of polymers and their classification, as well as the main domains of applications. Following, we illustrate the polymerization process through both step and addition methods. Sections 7.3, 7.4, and 7.5 highlight the most common polymer manufacturing techniques that dominate today's industrial production of polymer products: Injection Molding, Extrusion Molding, and Blow Molding. In the following "Problem" section, we detail composites manufacturing, whereas composites are typically a heterogeneous engineering material with both a matrix and reinforcement material components. The most common matrix materials for composites are polymers (thermosets) – more details on composites will be presented in the next chapter (Chapter 8).

FIGURE 7.1   Classic Mercedes Benz interior vs nowadays cars with an average of 260 lb/car.

7.1 **Polymers**

The term "polymers" originates from Greek origins, as it joins both "polus" and "meros." Polus stands for "many" and meros for "parts." Therefore, a polymer is understood as a collection of "many" monomers. Polymers cover a wide range of natural and synthetic materials. Natural polymers exist in both flora and fauna, including proteins, cellulose, and natural rubbers, among many others. Synthetic polymers are derived from carbon sources such as mineral oils, natural gas, and coal. They include epoxies, phenolics, polyethylene, and nylons.

The discovery of polymers was an accidental by-product of elephant endangerment. In the nineteenth century, billiard tables became very fashionable and billiard balls were made from ivory, harvested from elephant tusks. To produce 16 pool balls, unfortunately one to two elephants needed to die for the raw material. This lead a New York-based company to release a competition for scientists to find an alternative method/material to produce billiard balls replacing the omnipresent ivory billiard balls. Through participation in this competition, John Wesley Hyatt mixed pyroxylin and nitric acid with camphor to create celluloid. This enabled the Hyatt Brothers in the 1870s to invent the first polymer injection molding machine, and thus unlocking the evolution of polymers manufacturing. It is worthwhile to also note that Hyatt discovered polymers through participation in the competition; he ultimately did not win it! One reason was that those novel billiard balls, mainly made of celluloid, were highly combustible – leading to some "small explosions" that regularly happened when a player took a perfect shot at the pool table (Eschner 2017). Today, some polymers are still considered a fire hazard, and once enflamed are hard to extinguish. For example, we all have seen burning tires used as blockades during violent protests in the news.

Polymers' widespread usage came about for their highly desirable properties: operating temperature, friction, impact strength, cost, moisture absorption, compressive strength, and tensile strength. A main hindrance and adoption barrier is their lack of strength needed for some engineering applications; however that is only true for generic polymers. Recently, engineered thermoplastics such as PEEK and Polyethylenimine (PEI) are successfully used in high-end engineering applications in bearings, pumps, and other highly stressed applications. Another example for a special polymer from the PEI family, is ULTEM. The latter has well-known usage due to its heat resistance and especially its flame resistance. ULTEM 9085, part of the ULTEM family, is particularly suited for the aerospace industry and is considered as a flame-retardant. ULTEM is trademarked by SABIC Global Technologies B.V. Moreover, as polymers are resistant to corrosion, they possess electrical insulation aptitude and are typically a cheap option to consider.

Polymers can be split into three main groups: Thermosets, Thermoplastics, and Elastomers. Both Thermosets and Thermoplastics are considered part of the "Plastics" family. The split between the different polymers is aligned with their polymerization method (covered in more detail in the following section) used to create the polymer out of its base monomers. Indeed, the molecular structure of the polymer provides an indication of its polymer group. Elastomers are traditionally thermosets (such as rubber); however, there are also thermoplastic alternatives (such as polypropylene-based copolymer). We will discuss this in more detail in the respective sub-sections on thermosets, thermoplastics, and elastomers.

Today, the market share of polymers is clearly inclined towards thermoplastics. Actually, over 70% of the polymer production can be associated with thermoplastics, while both thermosets and elastomers jointly make up less than 30% of the polymers' market share. It is worthwhile to mention that, on a volumetric basis, annual usage of polymers trumps the annual usage of metals. The following sections present thermoplastic polymers, thermoset polymers, and elastomers (Figure 7.2).

7.1.1 **Thermoplastic Polymers**

Thermoplastic polymers are the most prominent form out of the main polymer groups, holding the biggest market share of all polymers. We can distinguish between traditional thermoplastics and engineering thermoplastics. The term engineering thermoplastics refers to superior properties in terms of strength and resistance to impact, heat, and chemical attacks – all highly desired for advanced engineering applications in industries such as aerospace and defense.

FIGURE 7.2 Polymers in the context of engineering materials.

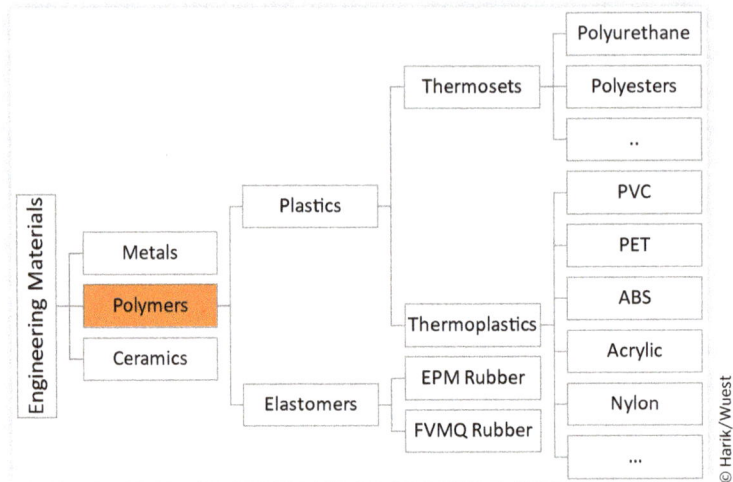

FIGURE 7.3 (Left) Water bottles are thermoplastics and you can easily spot the recycling number on them; (Right) Recycling codes.

© Mariyana M/Shutterstock; © CB studio/Shutterstock

The fundamental concept of thermoplastic polymers is that they can be subjected to multiple heating and cooling cycles with little to no degradation of the material properties. This makes them a more sustainable option, and thus highly desirable from an environmental standpoint as they can be recycled and reused. The term thermoplastic is the composite of "Thermal" and "Plastics" to emphasize the thermal effect and the ability to absorb thermal variations. Thermoplastics are typically brittle, and the addition of plasticizers and impact modifiers reduce their brittleness. Thermoplastics are defined by a linear or branched molecular structure that is comparably simple.

Thermoplastic, due to their ability to be reshaped and absorb heat, are highly recyclable, and as such, the Society of the Plastics Industry (SPI) created codes to help with recycling and sorting of different subgroups of thermoplastics. There are seven recycling codes depicted in this SPI categorization: (1) PETE for Polyethylene Terephthalate, (2) HDPE for High Density Polyethylene, (3) PVC for Polyvinyl Chloride, (4) LDPE for Low Density Polyethylene, (5) PP for Polypropylene, (6) PS for Polystyrene, and (7) grouping all other thermoplastics that do not fall within categories 1–6. It has to be noted, while thermoplastics can be recycled, their material properties degrade each time they are heated. An additional challenge recycling facilities face regularly is when the purity of the recycled material mix is low. The resulting products manufactured from fully or partially recycled materials have generally reduced material properties, which limits the extent of their practical recyclability (Figure 7.3).

Since 2018, the above recycling system is administered by *ASTM D7611—Standard Practice for Coding Plastic Manufactured Articles for Resin Identification.*

7.1.2 Thermoset Polymers

Thermoset polymers are "set" through a thermal process and reversing this process is not possible. Thermosets change irrevocably their chemical composition during the polymerization process. Thermosets have a tightly cross-linked molecular structure that is considered complex.

FIGURE 7.4   Prominent thermoset application: matrix material for composites.

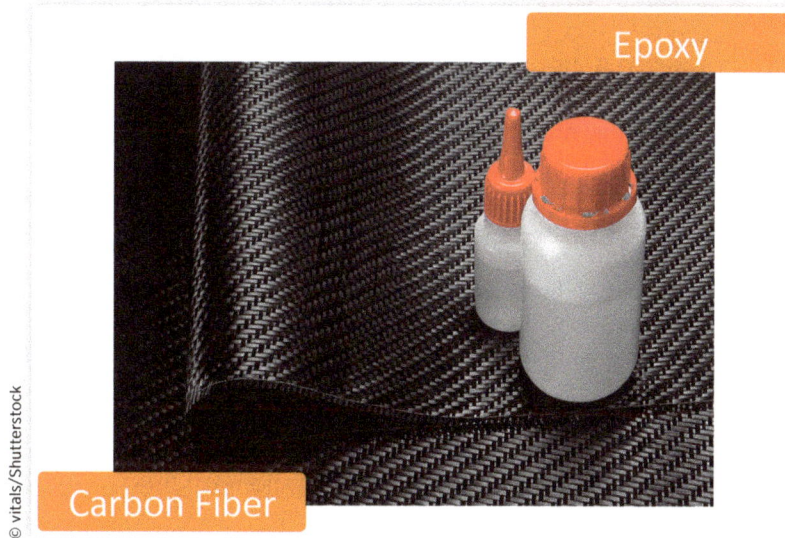

Epoxy

Carbon Fiber

© vitals/Shutterstock

Thermoset polymers are used as matrix materials in most composites applications (Figure 7.4). This setting is often identified as curing. Thermoset polymers are brittle and rigid, have a high range of operating temperature, and generally have a higher resistance to solvents. Thermosets, since they are "set" and cannot be subject to reheat/reshaping, are not recyclable. The curing process influences the molecular structure in a way that renders them unable to melt, but rather they would char or burn if subject to reheating. Many repurposing companies focus on chopping thermosets in small pieces for reuse as filler materials for other applications. In some cases, these fillers are used for new polymer products by mixing them with new polymer material. However, we need to be aware that the material properties of a part with such fillers suffer compared to using solely new material. Therefore, we need to carefully consider the require-ments before thinking about this option (Figure 7.4).

To better explain thermosets into their context, we present HexPly 8552 (https://www.hexcel. com/user_area/content_media/raw/HexPly_8552_us_DataSheet.pdf) as an example. HexPly 8552 is a mid-toughened, high-strength, damage-resistant, structural epoxy matrix known for its "350" cure cycle. The 350 refers to the temperature at which the polymerization is completed. Curing requires an autoclave capable of providing a combination of both heat and pressure. Figure 7.5 presents a cure cycle where we mesh the application of vacuum, pressure, and increase/maintain/decrease polymerization to achieve gelation and polymerization.

FIGURE 7.5   Cure cycle for HexPly8552 epoxy system.

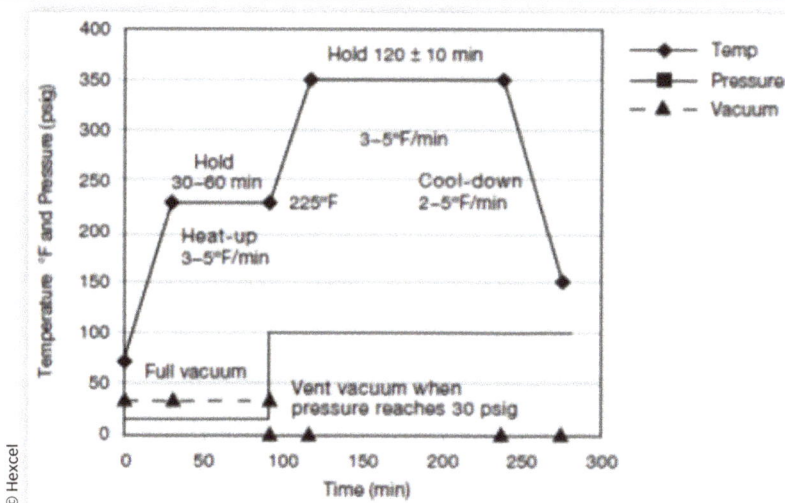

© Hexcel

7.1.3 **Elastomers**

Elastomers are elastic polymers that elongate when subjected to loads and return to their original form once loads are removed. They are capable for full recovery with little to no permanent deformation. Elastomers are defined by a loosely cross-linked molecular structure, considered to be comparably complex, that results in their advanced elongation capacity. Elastomers occur both in natural or synthetic form. The most prominent and widespread usage of elastomers is in tires for the automotive and trucking sector (Figure 7.6). Other applications include but are not limited to seals, O-rings, noise reduction, windshield wipers, and everyday items such as bumpers and children's toys.

Shore hardness is typically the measure of hardness for polymers in general, particularly useful for rubbers and elastomers. The higher the number in the scale the harder the material. Similarly, materials with low shore numbers are softer. To test shore hardness, we typically use a durometer which indents the material. There are several scales that covers the spectrum of material. For example, a chewing gum is scale OO 20, automated fiber placement rollers are A 70, and hard hats are D 75. The most common used shore hardness scales are A and D. Shore A is used for flexible materials, while D is used for harder materials. There is some overlap of the two scales.

7.1.4 **Polymer Properties**

In this section we discuss two different properties of polymers that are of importance for engineering applications and the polymers manufacturing process.

Glass Transition Temperature is the range where temperature enables the occurrence of glass transition. The latter is where polymers move (in both ways) between a viscous and a solid state. It is very important to remember that the pressure conditions influence this temperature range. We can use a Differential Scanning Calorimetry to perform thermal analysis on a sample and obtain more information on the thermal behavior of the polymer.

Tensile Strength: Similarly, to metals, we can test how much specimens can elongate prior to necking and thus value the tensile strength of a certain polymer material. The tensile strength generally increases with a higher degree of polymerization.

There are many material properties that influence which materials are used in engineering applications. They can include such properties as these two, but also include UV stability, notch sensitivity, water absorption, chemical resistance, stiffness, color fastness, hardness, plasticizer migration resistance, clarity, surface properties, and more. Processing is also a factor, as a more expensive material that processes faster can have lower production costs.

FIGURE 7.6   Most prominent elastomer: tires.

© Oleksndr_Delyk/Shutterstock

7.2 **Polymerization**

As described before, polymers consist of several (thousands) of monomers that are grouped together. This process of polymerization has an effect on the final polymer and its degree of polymerization. The latter is the number of polymers that were added or combined together to obtain the desired polymer. The degree of polymerization directly correlates with the final weight of the polymer.

It is very important to understand the connection between the polymerization method, the resulting molecular structure, and how this actually creates thermoplastics, thermosets, and elastomers. To that end, we first present a section where we introduce the two main polymerization methods step-polymerization and addition polymerization. Following, we present the different molecular structures of polymers and discuss the connection between their attributes and their molecular structure. One aspect we need to consider is that some polymers are created in a way that they have components that are still inactive, and as such they transform based on certain conditions from one type to another. An example is temperature activation that can occur during the use of a product, where a thermoplastic becomes a thermoset. We all experienced how a formerly elastic rubber band (elastomer) turned hard and brittle when left exposed to the sun for too long.

7.2.1 **Polymerization Methods**

Polymerization methods refer to the chemical process where we are linking the basic monomers to create the desired polymer. We have two main methods for achieving this polymerization: Addition and Step Polymerization.

Figure 7.7 shows addition polymerization, where a polymer of degree n-1 gets an additional similar monomer added – hence the addition label – to become of degree n. Figure 7.8 illustrates step polymerization, where two polymers of degrees' u and v, respectively, gets joined to become a polymer of degree n = u + v.

7.2.2 **Molecular Structures of Polymers**

The molecular structure of a polymer inherently explains the type of the polymer that is obtained by the polymerization process above. The macromolecular structure that is resulting from the chemical process can result into anything ranging from simple linear macrostructures to dense and cross-linked ones. The simpler the resulting structure the easier it is to reshape and recycle. The more complex the structure the harder it is to reshape and recycle the polymer material. We learned in the previous sections that thermoplastics can be easily recycled. Therefore, we can anticipate that the molecular structure of thermoplastics are comparably simple ones. Similarly, we understand that the molecular structure of thermosets is comparably complex and results in the distinct disadvantage of the polymer with respect to its recycling capability.

Next, we present the four main types of molecular structures in more detail: (1) Linear, (2) Linear Branched, (3) Loosely Cross-Linked, and (4) Tightly Cross-Linked. *Linear structures* are typically found in thermoplastic polymers. The linear structure provides fluidity for

FIGURE 7.7   Addition polymerization.

© Harik/Wuest

FIGURE 7.8 Step polymerization.

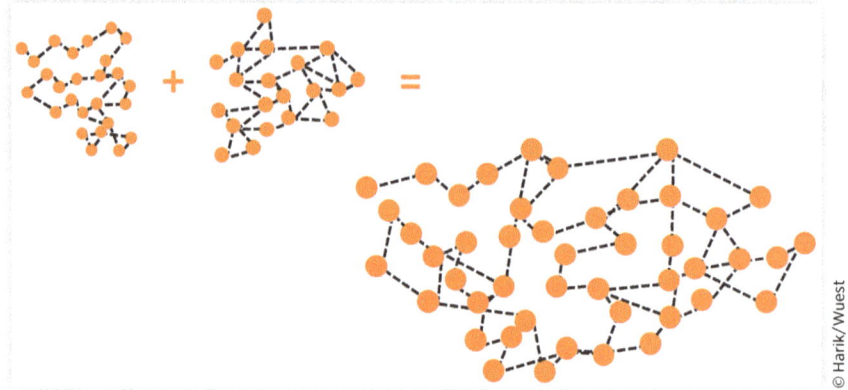

© Harik/Wuest

FIGURE 7.9 Linear (top) and linear-branched (bottom) molecular structures are a characteristic of thermoplastics.

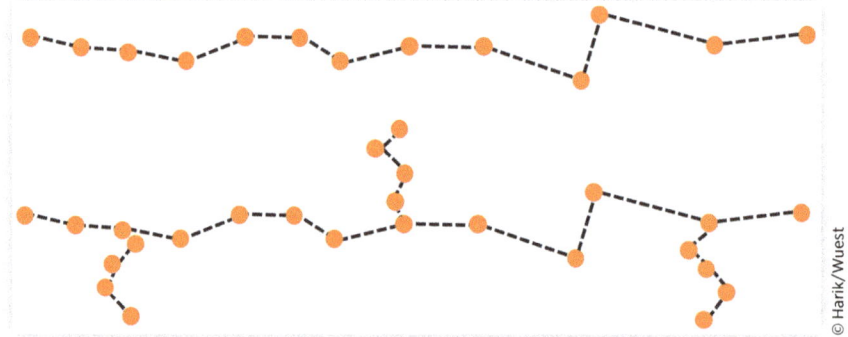

© Harik/Wuest

reheating/re-melting and as such enables the recycling of thermoplastics. Similarly, the *branched structure*, also considered a comparably complex structure, provides the ability to recycle and characterizes a material as a thermoplastic polymer. Cross-linked structures, whether loosely or tightly, are characterized by the absence of the ability of a polymer to be recycled. *Loosely cross-linked molecular structures* provide the advantage of deforming and returning (elastic deformation) to their original shape. Therefore, polymers sporting these molecular structures are defined as elastomers. *Tightly cross-linked molecular structures* are a characteristic of thermosets where the complex molecular bonds cannot be broken easily, and thus prevent the material from being recyclable once cured (Figures 7.9 and 7.10).

FIGURE 7.10 (Left) Loosely cross-linked structures are a characteristic of elastomers and (right) tightly cross-linked structures are a characteristic of elastomers.

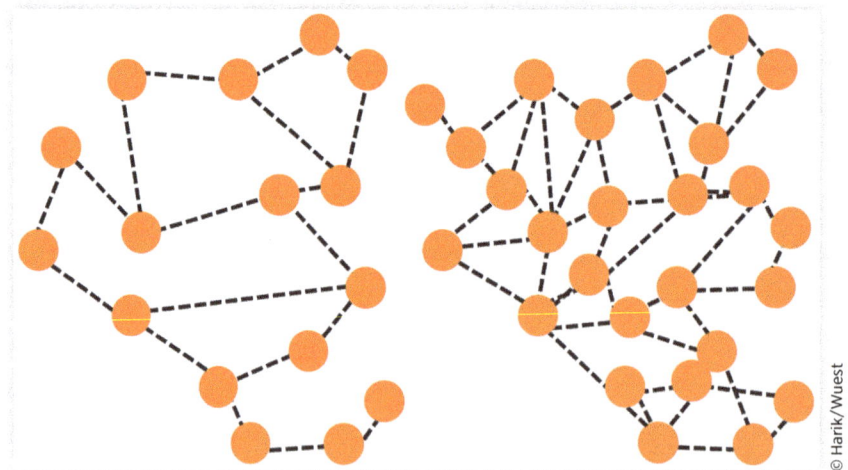

© Harik/Wuest

7.3 **Injection Molding**

We present Injection Molding, patented by the Hyatt Brothers in 1872, in the following three sub-sections. At first, we discuss the basic process setup and concept of an injection molding manufacturing process, followed by a selection of common applications. We conclude this section with a brief discussion of the distinct advantages and disadvantages of the injection molding process that generates 3D shapes.

7.3.1 **Process Concept**

Injection Molding is very similar to casting processes where we are pouring a molten metal into a die and wait for it to solidify. Injection molding describes the process of injecting molten polymer material under pressure into a mold cavity to solidify. Depending on the product or part size, we often group several parts or products together to be injection molded in one shot to increase the throughput of the process. Injection molding can accommodate multi-material and/or multicolor products in one shot; however, such a setup significantly increases planning efforts, tool design, and ultimately cost. Nevertheless, it is a commonly used option given a large enough production run.

The generalized injection molding process (Figure 7.11) proceeds with the following steps: (1) Preparation of Raw Materials, (2) Melting and Mixing of Materials, (3) Injection of Material, (4) Cooling of Materials, and (5) Ejection of Parts (Figure 7.12).

FIGURE 7.11 Generalized injection molding process.

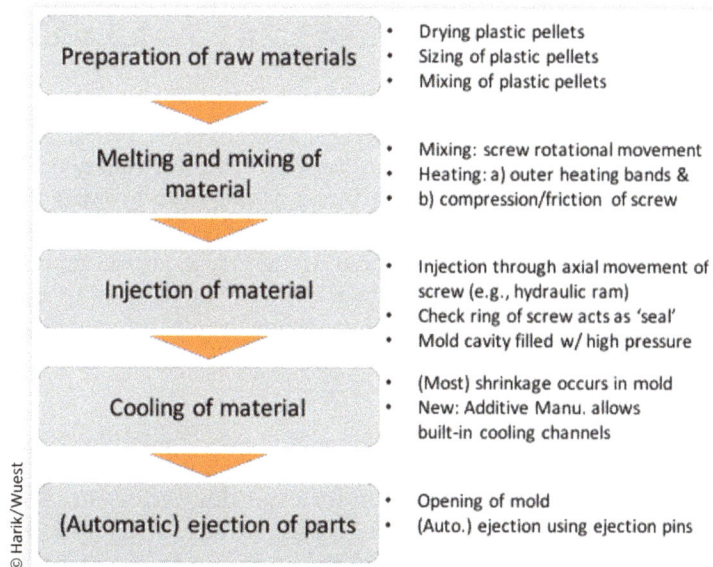

FIGURE 7.12 Process concept of injection molding.

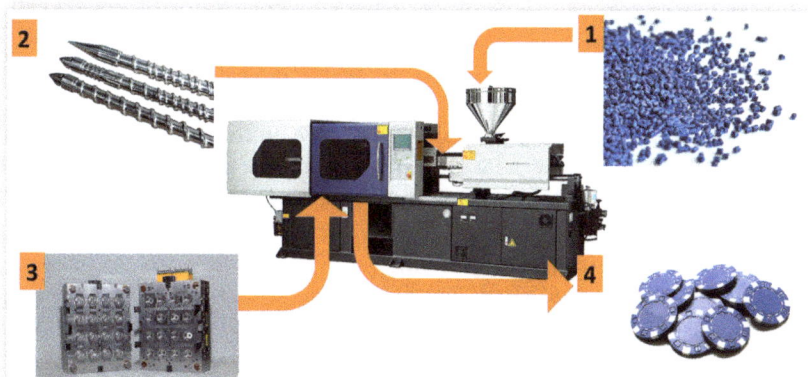

The first step of material preparation includes drying, sizing, and mixing of plastic pellets with other required materials for the process such as colorants or toughening agents. The second and third steps are typically obtained through the usage of a three-section screw, called reciprocating screw, that feeds, mixes, compresses, heats, and meters the polymer material. The reciprocating screw has the function, through its rotational movement and design, to heat, melt, and mix the materials with the goal of generating a uniform and homogenous material to be fed/injected into the mold. The design of the screw solves a key problem when it comes to achieving a uniform molding temperature of the polymer material - the generally bad heat transfer of polymers. Before the reciprocating screw was introduced, the polymers in an injection molding machine was very unevenly heated, leading to either a too high temperature close to the heating bands negatively impacting the material properties or too low temperatures at the inner core reducing the viscosity and as such causing problems during injection. The screw keeps the material close to the heating source and introduces heat through compression and friction during mixing, thus achieving homogeneous temperatures for higher quality molds. It can be safely said that without the invention of the reciprocating screw, modern polymers manufacturing would not be as omnipresent as it is today. The axial movement of the screw towards the mold chamber enables the material injection. The fourth step requires extra attention in order to avoid possible manufacturing defects such as voids and cracks in the final part. Finally, injection molding machine have ejector pins that supports the ejection of the part as soon as the part is solidified, and the mold is opened. To reduce the curing time, modern dies have cooling tunnels built in close to the mold surface to speed up the cooling of the part. Additive Manufacturing shows great promise to improve the manufacturing and use of complex incorporated cooling tunnels in injection molding dies.

Tooling and die design for injection molding is an important step and has significant impact on part quality and cost per part. High quality injection molding dies can produce several million parts before they wear out, but also can cost more than a million American dollars to make. On the other side of the spectrum, lower quality dies may only be good for 5,000 parts or less. However, low-quality dies, often made of softer aluminum, are significantly cheaper to design and make. Thus, the manufacturing engineer needs to make a judgment call on how many parts are to be produced overall, and then carefully calculate what option is preferable from a quality and cost standpoint. An option to increase flexibility is to design the molds with exchangeable inserts to enable the molding of varying parts using the same base die. An example where this is beneficial is a product where different springs are required to support changing loads. By using exchangeable inserts, the products can be manufactured using an injection molding process with n different diameter holes to accommodate the different diameter springs at significantly lower cost compared to using a new die each time.

7.3.2 Applications

There is a wide variation of applications using injection molding techniques. Several items of daily use such as combs, cups, plastic forks, and knives are made by injection molding. Toys such as toy cars and the famous Lego bricks are also manufactured using an injection molding process. Legos deserve our attention as they exemplify some very elegant solutions to injection molding process limitations. Injection molded parts are required to a slight draft angle for easier ejection and surface quality. However, Lego brinks require perfect 90° angles to function. The engineers solved this issue by adding the draft angle inside the brick where the 90° requirement does not apply. Two other aspects are parting lines and ejection pin marks. The tool and die design places the parting line directly at the bottom edge and thus makes it disappear on the final brick, and the ejection pin marks are well hidden on the bottom of the brick as well. Take one of your Lego bricks and have a look and find the ejection pin marks, internal draft angle, and see if you can see the parting line of the two die halves. Finally, industries heavily depend and rely on injection molding to produce needed parts for automotive, aerospace, medical and several other industries (Figure 7.13).

Injection molding is also often a first step in polymers manufacturing for other subsequent production methods. Preforms, made with injection molding, can be used in Blow Molding for the manufacturing of water bottles.

Applications include both durable goods (e.g., cars, appliances, cell phones, toys) and single-use items (e.g., cutlery, cups). Injection molded preforms are not used much anymore, but insert

FIGURE 7.13 Products made with injection molding.

© Yuliyan Velchev/Shutterstock; © Winai Tepsuttinun/Shutterstock; © Sever180/Shutterstock

molding is common, that is, a molded part can have metal components such as screw bosses integrated into them. Molded parts with two plastic materials are also common and can be made one of two ways. One way is overmolding, which is to mold a part in one type of plastic, then place it in a different machine to mold another material around it. The other is co-injection, where two materials are injected into the same mold. Common applications of insert molding include tools and toothbrushes.

7.3.3 Advantages and Disadvantages

Following we list the numerous advantages of injection molding, as well as we highlight the major disadvantages and deficiencies.

Injection molding advantages:

- Highly automated process
- High throughput (fast production)
- Capable of producing detailed features
- Capable of producing complex geometry
- Very efficient (given large batch size)
- Ability to enhance material strength using fillers
- Ability to create multi-material parts in one mold
- No to little post-processing required
- Reduced waste/scrap

Injection molding disadvantages:

- (Very) high tooling cost (mainly molds; partly special screws)
- There are certain restrictions regarding part designs (e.g., taper/draft angle, wall thickness, radii, and undercut)
- Only economically feasible for larger batch sizes

7.4 Extrusion Molding

We present Extrusion Molding structured in the following three sub-sections. At first, we discuss the process concept of extrusion molding, followed by common applications. We conclude again with advantages and disadvantages of the extrusion molding process that generated 2D profiles. Similar to injection molding, we can produce multi-material and/or multicolor products in an extrusion molding process. Combining rigid (thermoplastics) and elastic (elastomers or thermoplastic elastomers/TPE) can create highly desirable properties for specialty products, such as an AC curtain where the rigid material provides the structural integrity and the elastic parts in between the flexibility to open and close the curtain. Designing the dies (orifices) is very difficult and requires significant expertise. Often extruding experts prefer to work with materials they know and have worked with before, as new materials generally lead to some trial and error which is expensive and unproductive. This leads to slow adoption of advanced polymer materials in the industry. Materials for extrusion vary from the ones used in injection molding and are not interchangeable. Overall, there are also far fewer extrusion grades offered by material suppliers than there are materials for injection molding. Extrusion grades of materials must have much higher melt viscosity than injection grades as they need to have a higher structural integrity directly after exiting the orifice to not cause quality issues. Because sales volumes of extrusion grades are lower, material suppliers do not make as many different grades as they do for the injection molding market.

7.4.1 Process Concept

The extrusion molding process has different material grades required for other polymer manufacturing processes. Theoretically, since this is an extrusion process, we can produce endless parts. This is one of the main differences between polymers extrusion (theoretically endless – fed by polymer pellets) and metal extrusion (finite – fed by metal billet). In reality, shapes produced by polymers extrusion molding have finite length and are often with constant cross section. Profiles that can be extruded are often labeled by a letter name, as the extruded profile resembles it. We often find U-, O-, T-, and H-shaped as well as square-shaped extruded profiles. Also, we can manufacture very thin films and sheets. Thin films with a thickness of less than 0.02in and sheets that are less than 0.5in can be obtained, given a suitable polymer material is chosen. The manufacturing process is generally followed by a materials handling system such as a belt system. After the core extrusion process, we transport the material for further post-processing. Generated products are automatically cooled using a water bath, fan, or water spray. Parts are cut to desired length, for rigid or flexible material, or are coiled for flexible materials such as foil or garden hoses (Figure 7.14).

Below we describe the generalized extrusion molding process steps: (1) Preparation of Raw Materials, (2) Melting and Mixing of Material, (3) Extrusion of Material, (4) Cooling of Material, and (5) Cutting/Coiling (Figure 7.15).

In a first step, the preparation of raw materials (1) includes drying of the plastic pellets, followed by sizing and mixing. This ensures that the input material for the process is available in the right quality. Similar to other polymer manufacturing processes, material that is not dried can lead to quality issues of the final products such as void spaces caused as water turns to steam within the polymer material in the mold. This can cause surface holes and voids, and can furthermore cause dangerous overpressure problems.

Next, we use a screw rotation to perform the second step of melting and mixing of material (2). In this step, the screw rotation acts as both the mixing and transportation element. Different

FIGURE 7.14 Extrusion molding of thin bags.

FIGURE 7.15 Generalized extrusion molding process.

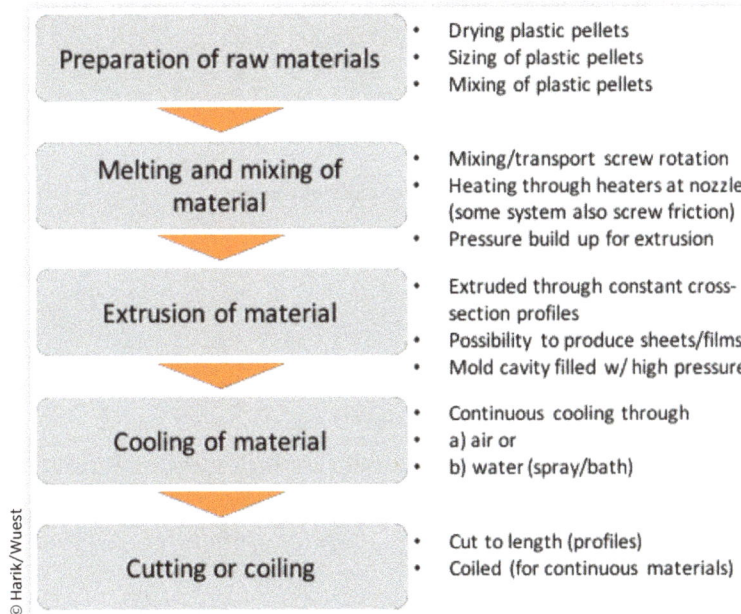

Preparation of raw materials	• Drying plastic pellets
	• Sizing of plastic pellets
	• Mixing of plastic pellets

Melting and mixing of material	• Mixing/transport screw rotation
	• Heating through heaters at nozzle (some system also screw friction)
	• Pressure build up for extrusion

Extrusion of material	• Extruded through constant cross-section profiles
	• Possibility to produce sheets/films
	• Mold cavity filled w/ high pressure

Cooling of material	• Continuous cooling through
	• a) air or
	• b) water (spray/bath)

| Cutting or coiling | • Cut to length (profiles) |
| | • Coiled (for continuous materials) |

© Harik/Wuest

from the injection molding process and its screw, the extrusion process is continuous and not supported by a ram. The heating through heaters at nozzles, combined with the pressure build up, ensure adequate forces to perform the next step. The third step of material extrusion (3) is achieved by the pressure build-up pushing the heated polymer material through the die. Following the material extrusion, cooling (4) is performed by means of air, bath, and/or a spray of water. It is more common to allow cooling by water or water spray. The cooling is almost always aided by the use of fixturing to hold the shape until it is solid enough to hold itself in industrial practice.

Finally, we process to cut or coil (5) the product depending on the profiles or in the event we are producing continuous materials. For special applications there are many specialized tools available to support the final steps of the extrusion process. For example, the special case of extrusion molding films, such as used for trash bags, require specialized cooling fixtures that pull and stretch the material into the thin film during the process.

7.4.2 Applications

Extrusion processes generate a wide variety of constant cross section of products and parts. The most prominent example is plastic tubing and PVC pipes. Other examples are insulation for wires, rods, rails, seals, sheets, films, and also vinyl house siding. We can correlation the extrusion molding process to the Deformative Manufacturing extrusion process for metals covered in Chapter 2. Both durable goods such as garden hose, furniture edging, and underground piping are extruded, as well as disposable goods such as straws and trash bags (Figure 7.16).

FIGURE 7.16 Major applications of extrusion molding.

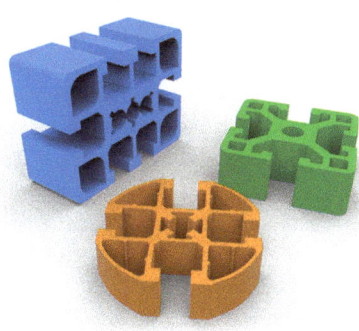

© NOOCHICO/Shutterstock; © Peter Sobolev/Shutterstock

7.4.3 **Advantages and Disadvantages**

Following, we list the numerous advantages of extrusion molding, as well as we highlight the major disadvantages and deficiencies.

Extrusion molding advantages:

- Comparably low tooling cost
- Ability to produce continuous parts (theoretically endless)
- Very high production volume (high throughput)
- Low cost of parts
- Efficient process
- Little to no waste/scrap
- Production of multi-material products possible
- Easy post-process manipulation (part comes out hot which allows certain alterations)

Extrusion molding disadvantages:

- Limited complexity of parts
- Limited to parts with constant cross section
- Limited material selection (compared to injection molding)

7.5 **Blow Molding**

We present blow molding through the following three sub-sections. At first, we discuss the process concept of blow molding, followed by applications in a second sub-section. We end this section with advantages and disadvantages of blow molding processes that generated 3D shapes.

7.5.1 **Process Concept**

Materials needed for blow molding is typically an injection-molding output that has a solid bottom and hollow preform. One can imagine that the blow molding process is always following a previous injection molding or extrusion molding process. Although injection molding outputs in form of so-called preforms is the typical input for the blow molding process of, e.g., plastic soda bottles, open-ended preforms obtained through an extrusion molding process are also commonly used. Many blow molding processes utilize an extruded tube (called a "parison") that is immediately clamped in the blow mold and formed, all in one process. Extrusion blow molding becomes basically a post process of extrusion in this case.

The manufactured shapes are typically seamless, hollow-shaped containers such as plastic bottles and car gas tanks. Good dimensional tolerances are achieved and the resulting part is easily removed from molds. Therefore, no ejection pins are required in a blow molding process setup. It is worthwhile to note that the process is available for both axial and radial expansion through the blown air burst. Multilayer products are also possible (Figure 7.17).

Below we describe the process of blow molding following an initial step where raw materials are prepared through a prior process: Injection Molding or Extrusion Molding. The blow molding process is performed through four steps: (1) Heating of Preform, (2) Mold Closure, (3) Blow Molding, and (4) Cooling/Ejection (Figure 7.18).

With respect to the heating of preform step (1), we increment the heat of the preform to the required temperature as per the specifications of the material. In this step, the preform is installed on the blow pin through which the air will burst. Next, we close the mold (2) where the heated preform will be trapped within the mold sides. At this stage we have void between the preform and the mold. Next, air pressure burst expands the preform against the sides of

FIGURE 7.17 Process concept of blow molding.

© Anton Starikov/Shutterstock; © Aumm graphixphoto/Shutterstock; © Travelerpix/Shutterstock;
© Christos Theologou/Shutterstock; © Travelerpix/Shutterstock

FIGURE 7.18 Generalized blow molding process.

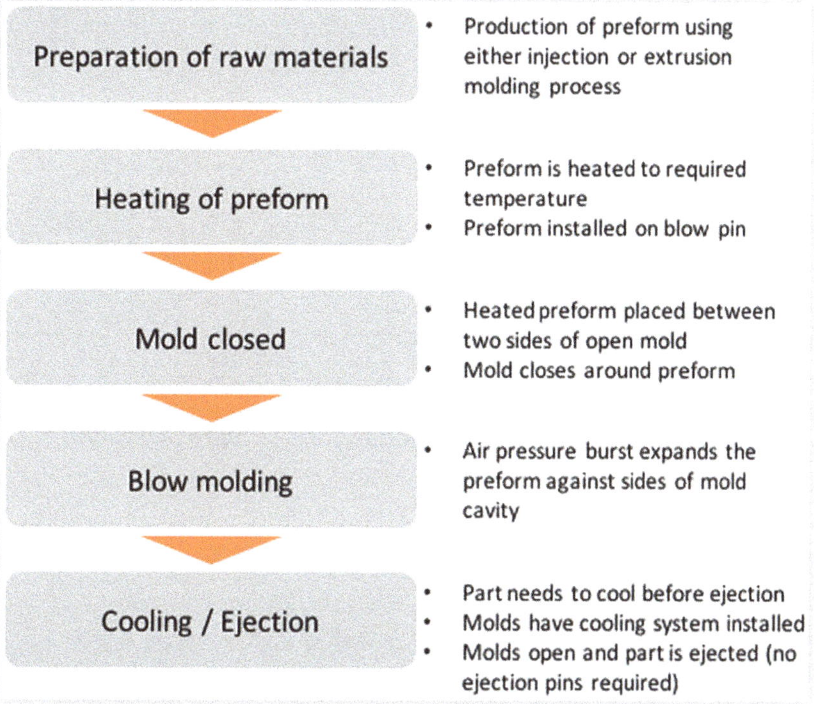

the mold cavity through the blow molding step (3). Finally, the parts need to cool before ejection. As soon as the molds are opened the part dropped and ejector pins are usually not needed in this final step of cooling/ejection (4).

7.5.2 Applications

Blow molding is mainly used to create products that have a hollow area (that is produced by the blow) to contain material. Everyday items such as water bottles and bottles, in general, gardening equipment, household items, and containers of all sorts are a direct application of Blow Molding. In the industrial domains containers and coolers are also blow-molded parts.

As stated in the injection molding section, blow molding requires a preform that is mostly a result of an injection molding process (Figure 7.19).

FIGURE 7.19 Major applications of blow molding processes.

© Christos Theologou/Shutterstock; © Oleksndr_Delyk/Shutterstock; © Brooke Becker/Shutterstock

7.5.3 Advantages and Disadvantages

Following, we list the numerous advantages of blow molding, as well as we highlight the major disadvantages and deficiencies.

Blow molding advantages:

- Comparably low mold cost (compared to injection molding) and system cost
- Possibility to mold (relatively) complex geometries (e.g., external threads)
- High production cycle/volume
- Automation level
- Ability to make hollow parts
- One-piece (seamless) products
- Multilayer products possible
- Excellent pressure performance of final parts

Blow molding disadvantages:

- Limited features (e.g., no holes, other than the opening for the blow pin, can be molded in)
- Wall thickness limited
- Limited to hollow containers of some sort
- Limitations on diameter of parts (to uphold tolerances in, e.g., corners)
- Comparably high amount of scrap parts (compared to extrusion/injection molding)
- Impact of products on the environment

Reference

Eschner, K., "Once Upon a Time, Exploding Billiard Balls Were an Everyday Thing," *Smithsonian*, 2017, retrieved from: https://www.smithsonianmag.com/smart-news/once-upon-time-exploding-billiard-balls-were-everyday-thing-180962751/.

Composites Manufacturing

Composite materials are heterogeneous combinations of two or more constituents with different mechanical, physical, and/or structural properties. The fundamental concept behind composite material is that the targeted combination of the different constituents provides superior properties for a specific engineering application compared to the individual constituents' performance. Composites are not a new development and are actually quite common in nature. Materials such as wood and bone can be classified as composites and have been arguably around for millions of years.

There are many forms of composite materials. However, typically a composite contains two main components: i) a reinforcement material which provides stiffness and/or strength and ii) a matrix material which surrounds the reinforcement component and holds the composite in its final shape. It has to be noted while in most applications two major components make up the composite material, there are applications for composites with three or more components can be that are combined to achieve the desired properties. Reinforcement components come in a variety of forms and shapes, from short and/or small particles to long continuous ones. The latter is usually the most desired variation from an end-user's perspective since this option offers excellent load-carrying properties, often resulting in highly desirable strength-to-weight and/or stiffness-to-weight ratios. From the perspective of the manufacturing engineer, however, the former variation, sporting short/small particles, is generally considered more desirable as it significantly reduces the complexity of design and manufacturing.

Today, fiberglass is one of the most frequently used composite materials.In fiberglass composites, the reinforcement is made out of continuous glass fibers while the matrix is made out of a resin system. In the aerospace world, carbon fiber reinforced polymers (CFRP) are one of the most advanced engineering materials, possessing highly desirable properties when designed and manufactured correctly. The high strength and high stiffness provided by the carbon fibers (reinforcement component) provides un-matched properties. Thermoset polymers are most commonly used as the matrix component today; however, thermoplastic polymers are slowly emerging as an alternative matrix system for CFRP. As we learned in the previous Chapter 7 "Polymers Manufacturing," recyclability is an unresolved issue for many composite materials. This switch towards thermoplastics as matrix materials might make composites even more appealing in the future as it would improve recyclability.

Numerous historical examples of composite products can be used to illustrate the fundamental concept and highlight the differences to traditional, non-composite engineering materials. In civil construction, adobe (in Spanish) is a brick made from mud as the matrix material and straw as the reinforcement material. These bricks are used in a three-layer construction structure including a first layer of adobe, followed by layers of mud and limestone. These homes were known for this "composite" stone, enabling them to be one of the most comfortable dwellings as they provided cool temperatures in the summer and warm temperatures in the winter. This example supports the fundamental concept that composites enable and unlock superior properties than the individual elements by themselves.

Composite products are widespread across multiple industrial domains, and their usage has been on the rise since the 1960s. While the aerospace world has been a pioneer in the adoption and dissemination of composites and composite products, several industries are pushing the boundaries of composites design and manufacturing today. Composites are utilized in many sporting products, such as golf clubs, long before they took off in aircrafts, since people were willing to pay top dollar to get a slightly better golf club or tennis racket. Additionally, if a club breaks, the repercussions are not as catastrophic as failure of a critical aircraft part, reducing the risk of early adoption significantly. In addition, boat hulls with high porosity have been around for quite a while since weight is not an issue. The novelty in the use of composites is to obtain quality aerospace parts that are also lightweight, reliable, and consistent. Today, it is very common to use composites in energy, automotive, aerospace, sports, biomedical, and a multitude of other industries. Perhaps the most common and well-known complex lightweight structures using composite materials are the Boeing 787 "Dreamliner" and the Airbus 350. It is estimated that composites constitute at least 50% by weight of the Boeing 787, whereas the older Boeing 777 uses only 12% of composites. In addition, many of the composite parts in the 777 are nonstructural parts such as lavatory panels, while in the new 787, the composite components are structural and load bearing. Boeing, Airbus, and several other manufacturers used the opportunity to learn and gain experience in composites design and manufacturing by integrating these nonstructural parts in the planes with the clear vision to utilize their superior properties in later designs throughout the whole airplane.

The use of composite materials can have a significant impact on the performance of the final product. While today the use of composites is often associated with additional complexity and cost in the design and manufacturing processes, the benefits during the operation of the product can outweigh these drawbacks. In the future, with improvements in our understanding composite manufacturing technologies and refinements in design, more advancements can be made. Using the example of the Boeing 787 and 777 as a comparison, we notice the increase in usage of composites of 50%. This not only reduces the overall weight of the aircraft, but also provides longer maintenance cycles, as well a better flying experience.

Textiles and textile manufacturing are often seen as precursors to composite manufacturing. It is quite common for composites manufactures to have their roots in the textile industry. Toray industries, for example, today one of the major companies in the composite materials market, had its roots in textiles and fabrics. Recently, Toray acquired TenCate, another company whose roots are also in the textile industry. Other major composite materials manufacturers include Hexcel and Solvay. Toray, Hexcel, and Solvay dominate the composite materials industry and manufacture a wide range of woven fabrics, tape, slit tape (for Automated Fiber Placement [AFP]), adhesives, reinforcements, honeycomb structures, resins, and other composite constituents or products.

This chapter provides an overview of composite manufacturing with a focus on AFP as it is used in the aerospace industry today. In order to provide a solid foundation for the readers, we start with a general overview including an introduction to composites terminology in Section 8.1. This first section describes the different types of composite materials, the concept of the stacking sequence and several other foundational parameters required to understand composite manufacturing techniques and principles. Section 8.2 delves deeper into the material itself and presents the micro-mechanics of composites. Within this section, we demonstrate both the directional properties of the fibers and introduce the volume fraction and rules of mixtures concepts. Composite manufacturing is primarily about achieving high fiber fraction volumes while keeping voids to a minimum. Understanding the basics of micro-mechanics paves the way to a better understanding of these concepts and the ability to apply them as a design or manufacturing engineer. In Section 8.3, we present several composite manufacturing techniques such as hand layup, pultrusion, resin transfer molding (RTM), and Automated Tape Laying (ATL). The latter is the precursor to AFP which is detailed in the final section (Section 8.4).

8.1 **Composites Terminology**

The distinct components of a composite material are typically identified as

1. *Reinforcement*: providing the strength component to the final material

2. *Matrix*: providing the shape/adhesion component

The reinforcement element, as stated earlier, can be continuous fibers or chopped fibers (see Figure 8.1). With respect to manufacturing, continuous fibers are more demanding and require additional control and physical mechanisms to ensure appropriate manufacturing compared to chopped fiber. There are also limited applications of aligned chopped fibers and long (several inches) discontinuous fibers. To illustrate and provide a better example, let us consider three variations of the fused filament fabrication (FFF) manufacturing process technique (compare Chapter 4 "Additive Manufacturing"). The first FFF process we consider is the most common variation, where the nozzle simply extrudes the thermoplastic material and it consolidates on contact to create the shape, layer by layer. The extrusion process in this case has exclusively one material (e.g., ABS, ULTEM) and has no distinct phases (homogeneous) and is, therefore, not a composite material. The second process variation includes injecting the original polymer material with chopped particles (e.g., short carbon fibers). During the extrusion process, the nozzle is delivering both materials, in their distinct phases, to create the final structure. This can also be achieved with pre-prepared filaments that already include the chopped fibers within the polymer or by mixing a single component filament at the extruder stage with chopped fibers. This second process does not require any further modification of the delivery system but rather some minor process parameter modifications that do not necessitate major machine adjustment. Let us now consider a third, more complex process variation, where the nozzle has to extrude simultaneously a continuous filament (e.g., continuous graphite fiber) at the same time as providing the surrounding matrix material (e.g., ULTEM). This final method requires significant machine adjustments, and one cannot simply use the traditional mechanisms of FFF 3D printing. Several additional steps and capacities are required to enable the process to deliver the desired product properties, such as cutting and nozzle design adjustment (Zhang et al., 2016).

Although the above examples all specified polymers as the matrix element and carbon as the reinforcement, this particular combination is not an exclusive combination to form composites. Any engineering material can be used for either matrix or reinforcement. Table 8.1 provides an overview of the different possible material combinations. Although some of these combinations are very common for a variety of reasons, including difficulty of manufacturing and/or cost, and some of them dominate the composite industry, it has to be noted that composites can always be created as long as each component is maintained in its distinct phase. However, the main question is what variations are most desirable from a functional, sustainability, and/or economics perspective (Table 8.2).

FIGURE 8.1 Continuous filament and chopped fibers (flake/particle).

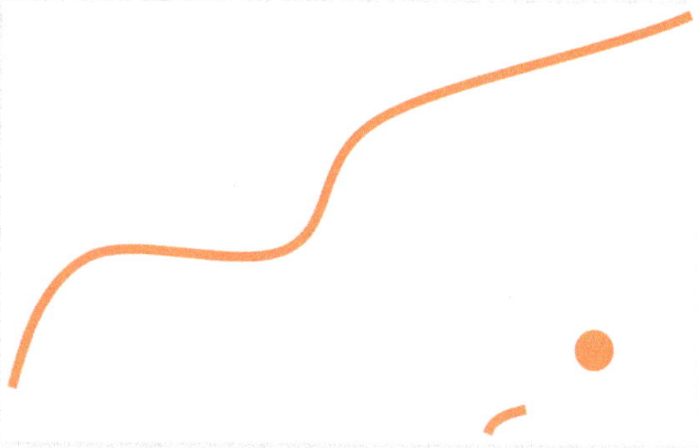

© Harik/Wuest

TABLE 8.1 Key characteristics of Boeing's 777 and 787 Dreamliner, according to http://www.modernairliners.com

Characteristic	777 300ER	787 - 8
Weight (Empty)	167,800 kg	119, 950 kg
Maximum Range	7,930 NM	7,355 NM
Passenger capacity	451 (2 class)	242 (2 class)
Maximum Fuel capacity	181,283 Liters	126,206 Liters
Composites	12%	50%
Aluminum	50%	20%

© Harik/Wuest

TABLE 8.2 Overview of possible reinforcement/matrix material combinations

		Reinforcement		
		Metal	**Ceramic**	**Polymer**
Matrix	Metal	Metal-metal composites	Metal-ceramic composites	Metal-polymer composites
	Ceramic	Ceramic-metal composites	Ceramic-ceramic composites	Ceramic-polymer composites
	Polymer	Polymer-metal composites	Polymer-ceramic composites	Polymer-polymer composites

© Harik/Wuest

8.1.1 Material Forms

Many material forms that are used to manufacture composite parts are available to the design and manufacturing engineer. In the following, we list the most common variations and briefly explain their predominant main applications.

Unidirectional Tape (see Figure 8.2): When unidirectional tape is used as an input material for composites manufacturing, all fibers within the tape layer have the same orientation. They are held together by the prepreg (matrix) material. An example of one of the most common unidirectional tapes is IM7-8552. This designation can be used to provide insights on the different components. Typically a carbon-epoxy material is identified first by the fiber then by the matrix. IM7 stands for "Intermediate Modulus" (IM) carbon fiber reinforcement material and 8552 is a thermoset resin system. It is very important to understand the resin system and its requirements (such as whether or not it is formulated to require an autoclave cure cycle). It is important to be aware that unidirectional tapes are not all necessarily available in form of prepreg composites. Recently a new wave of thermoplastic resins (in contrast to the established thermoset resins), dry tape, or low-tack tape has been introduced for industrial application. Several advantages motivated these innovations, predominately the need for out-of-autoclave process cycles. The latter concept will be illustrated in more detail in the AFP section.

Slit Tape (see Figures 8.3 and 8.4): Certain processes require the unidirectional tape described above to be slit into smaller dimensions that can be used for a specific manufacturing

FIGURE 8.2 Unidirectional tape with the orange strips representing the fiber orientation.

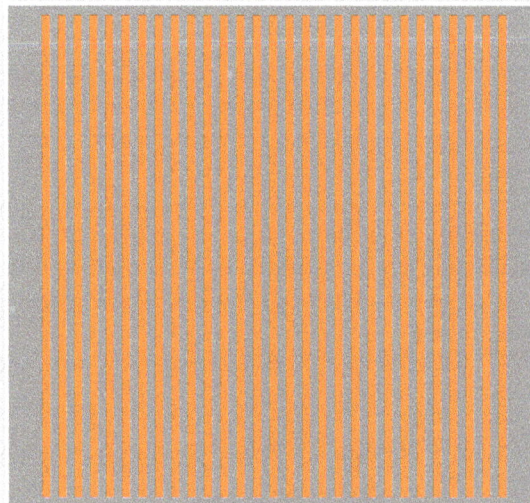

© Harik/Wuest

FIGURE 8.3 Slit tape of finite width depending on application.

© Harik/Wuest

process. Reasons include the creation of more intricate features and dimensions of machinery required. The fundamental difference between unidirectional tape and slit tape resembles the difference between the associated processes ATL (unidirectional tape) and AFP (slit tape). The dimensions of slit tape are typically resembling 1/8, ¼, ½, and 1 inch in width. Unidirectional tapes generally have widths of 3 inches and more.

FIGURE 8.4 Example of slit tapes rolled on spools.

© Andrew Anderson

Woven Fabric (see Figure 8.5): Woven fabric is one of the most commonly used formats that is mainly used for hand/manual layup techniques in composite manufacturing. As the terminology already indicates, woven fabrics are very similar to materials used in textiles and clothes. Woven fabric typically integrates fibers oriented in two or more directions in the same ply, woven into each other. Woven fabrics do not necessarily require prepreg material and are often used as dry fibers. The resin system is applied in a second step to fill the spaces between the fibers of the woven fabrics to create the intended shape.

Non-crimp Fabrics: To facilitate the use of composites, pre-forms that include several orientations of the in-plane fibers (reinforcement) are being created where the fibers are not woven together. Instead, the layers of dry unidirectional fibers are stacked on top of one another and held together by a thin thread through the thickness. Theoretically, fibers of any orientation can be used in this arrangement. This grouping of fibers is created on a machine and provides an intermediate material, thereby reducing the touch labor required to place individual plies by the number of plies in each "stack" of material. This approach can also increase the quality by reducing human error which can cause fiber misalignment and/or missing layers.

Resin Systems: There are several variations when it comes to resin systems. Today the aerospace industry predominately uses thermoset resin systems in aircraft primary structures. However, thermoplastics are likely to come into widespread use as soon as they are available at a competitive price and quality. Thermoplastics have a higher viscosity and, therefore, generally require higher heating and pressure cycles to achieve solidification compared to thermoset resin systems. For any resin system, one of the most important elements influencing in part quality that must be considered during the selection of a suitable resin system is the use of a proper cure cycle in terms of both temperature and pressure and the appropriate time constraints. Thermoplastics have the advantage of an out-of-autoclave process, hence their desirability despite their higher heating requirements.

FIGURE 8.5 Woven fabrics where we have two or more orientations in the same layer.

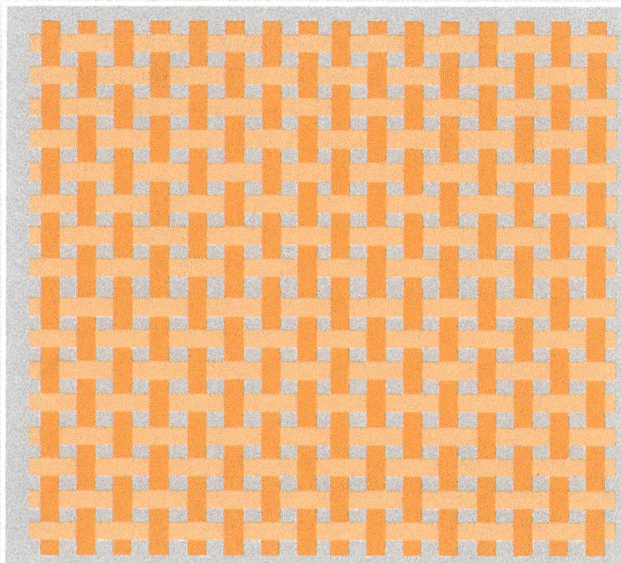

© Harik/Wuest

Adhesives: Fibers break during cutting and drilling of composite structures. This leads to delaminations or other defects in the final part which will reduce its strength. Several techniques can be used in some cases to eliminate, or at least reduce, the need for drilling. New techniques are being developed such as induction welding and adhesive bonding. These may replace some of the traditional mechanical joining techniques, such as the use of threaded fasteners, with more composites-friendly ones. One example of an adhesive that can be used to avoid drilling holes and adding fasteners through composite structures is Solvay's Metlbond 1515-3m. However, adhesives require adequate surface preparation using grit blasting or plasma treatment which limits their practical use.

8.1.2 Stacking Sequence

This section discusses what constitutes a "composite part" which is built from continuous fibers. In principle, we refer to a composite part as a collection of multiple plies stacked on top of each other in a layer-by-layer fashion. The creation of these plies is discussed later in this chapter. In this section, we present the stacking sequence and additional annotations that support this explanation.

FIGURE 8.6 Traditional layup orientations 0, +45, −45 and 90.

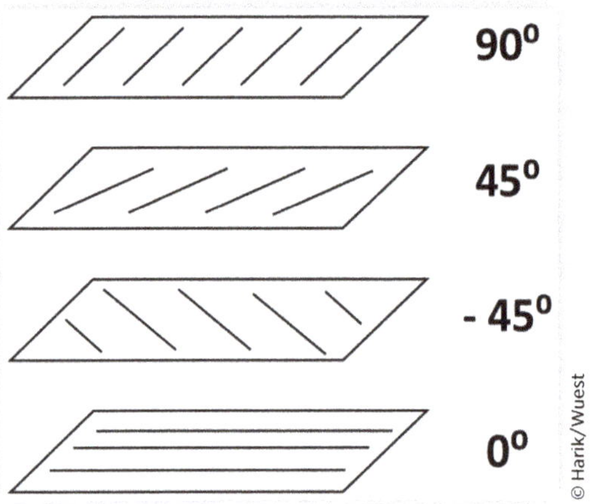

© Harik/Wuest

FIGURE 8.7 Balanced symmetrical quasi-isotropic composite laminate.

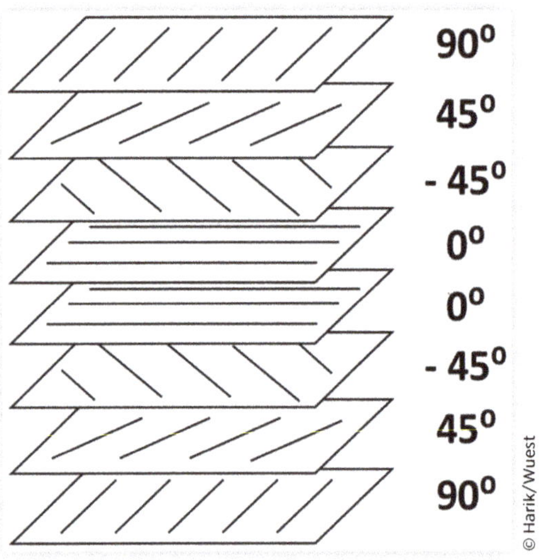

© Harik/Wuest

First, let us introduce the essential concept of angle and orientation in the context of fiber placement. Any unidirectional tape has its "along the fibers" orientation. A single layer of a unidirectional tape is referred to as a ply or laminate. These layers can be used to create a laminate where each ply may be oriented in a different direction. To define the ply orientations of the laminate, we set a reference orientation of 0° as the orientation along the longest side of the panel (or complex shaped part), or alternatively, as selected by design to achieve a specific property of the composite part. Subsequently, any other ply is labeled with respect to this original reference orientation. If the "along the fibers" orientation of a new ply deviates at a 45° angle from the reference orientation, this ply is classified as a 45° ply. Eventually, all individual plies are classified based on their angle with respect to the reference orientation. In the example below, the reference orientation is the bottom ply at 0° and all other plies are defined based on that orientation (Figure 8.6).

At times, we insert padding, spacers, or local reinforcement such as honeycomb layers into complex structures between fibrous plies. Moreover, with recent advances enabled by modern manufacturing techniques, individual plies can have orientations which change within the plane of the ply, resulting in panels with variable in-plane stiffnesses. This in-plane change in orientation enables improved tailoring for specific locations in a structure where the loading and strength/weight requirements are well defined. Designing and manufacturing such parts is not trivial and only economically feasible at this point for specialty products, such as in aerospace or military applications.

We define the stacking sequence by listing the orientations from the first layer to the last in sequence. The stacking sequence for the composite example below is first defined as [90, 45, −45, 0, 0, −45, 45, 90]. This nomenclature can be simplified by taking advantage of specific characteristics. If a composite laminate is symmetrical, the subscript s can be used to reduce the annotation. In the given example, since the composite is symmetrical around its centerline, it can be annotated as $[90, 45, -45, 0]_s$. An additional simplification in the notation can be employed when the same orientation occurs in the sequence for adjacent plies but with the opposite sign. In this case the notation for two plies can be merged, by introducing the ± sign (Figure 8.7). The annotation is then $[90, \pm45, 0]_s$.

Laminates are considered "balanced" when the laminate contains only pairs of plies at each orientation, i.e., for each +theta ply, there must be a −theta ply. The laminate above includes a −45° ply for each +45° ply and is therefore considered balanced.

If the composite panel includes repetition, we can use parenthesis and numerical subscripts to indicate this repetition. As such, […, 90, 0, 90, 0, 90, 0, …] becomes […, (90,0)$_3$, …]. In case the composite includes an asymmetrical middle layer, we can include a bar above that layer to indicate it is not repeated. As such, [90, 0, 45, 0, 90] becomes [90, 0, $\overline{45}$]$_s$. Additionally, while not required, sometimes a subscript "T" is added to the end of a stacking sequence to enforce the idea that the entire (total) laminate is defined so that the reader does not question whether an "s" subscript was merely forgotten.

8.2 Micro-Mechanics of Composite Structures

This section illustrates the different directional properties of composites. As previously stated, composites are anisotropic, and it is extremely important to understand the orientation of the material. In comparison, when designing a part using metals, one only has to consider strength, weight, and volume. These features are sufficient to formulate the design problem and attempt to solve it. For composites, several additional factors come in play that significantly affect the properties of the final structure. Therefore, design and manufacturing engineers need to consider these factors. For example, a flat panel with an unbalanced or unsymmetrical stacking sequence will warp when cured and loaded. Careful attention to fiber orientations is necessary to ensure that a suitable, composite-specific design is developed that is also manufacturable. One of the most common deficiencies within the composites' design world is that designers who's main previous experience is with metals must adjust their mentality to work with composites. It is no longer possible to propose a design without a solid understanding of the manufacturing techniques and their influence on the behavior of the final part.

8.2.1 Theory

Composite manufacturing techniques are geared towards the increase of the fiber volume while achieving adequate cohesion. To best illustrate this concept, we will annotate data relevant to the reinforcement phase by **r**, to the matrix phase by **m**, and to the overall composite by **c**. Manufacturing techniques aspire to eliminate air bubbles and voids trapped within a laminate. Unfortunately, such defects cannot always be avoided. The prevalence of these defects depends on the resin material and the resin curing process. Autoclaves, employing a cycle of pressure and heat, are geared towards successfully reducing voids to a non-meaningful portion (<1% of overall volume). In general, we account for the volume of voids, using the subscript **v**.

The composite total mass (m) and volume (V) are then computed as

$$m_c = m_r + m_m + m_v \qquad \text{Eq. (8.1)}$$

$$V_c = V_r + V_m + V_v \qquad \text{Eq. (8.2)}$$

If an advanced manufacturing technique is properly used, such as an autoclave or an innovative Out-of-Autoclave process, we can update the above equations by assuming the void volume is less than 1%, leading to the following simplified equation:

$$m_c = m_r + m_m \qquad \text{Eq. (8.3)}$$

$$V_c = V_r + V_m \qquad \text{Eq. (8.4)}$$

CHAPTER 8

8.2.2 Volume Fractions

We now introduce the concept of the volume fractions defined by the volumetric ratio of individual constituents with respect to the total volume. Designers aspire for the highest fiber volume structure possible while ensuring the matrix material is doing its job: providing appropriate cohesion to the fibers. Considering the above definition, we compute the volume fractions, annotated by f, as follows:

$$f_m = \frac{V_m}{V_c} \qquad \text{Eq. (8.5)}$$

$$f_r = \frac{V_r}{V_c} \qquad \text{Eq. (8.6)}$$

Considering the prior condition of a "no void" (= 1% void space) composite, the volume fractions can be expressed as

$$f_m + f_r = \frac{V_m}{V_c} + \frac{V_r}{V_c} = \frac{V_m + V_r}{V_c} = \frac{V_c}{V_c} = 1 \qquad \text{Eq. (8.7)}$$

8.2.3 Density of Composite Structure

The density of a composite part ρ_c can be calculated using the rules of mixtures. Below we prove the direct application of the rules of mixtures and present a small application example.

$$\rho_c = \frac{m_c}{V_c} = \frac{m_r + m_m}{V_c} = \frac{\rho_r \times V_r + \rho_m \times V_m}{V_c} = \frac{\rho_r \times V_r}{V_c} + \frac{\rho_m \times V_m}{V_c} = \rho_r \times f_r + \rho_m \times f_m \qquad \text{Eq. (8.8)}$$

A composite structure made with IM7/8552 has 40% volume of fibers. To compute the density of the structure, we first retrieve the individual densities of the IM7 reinforcement and the 8552 matrix system. IM7 by Hexcel, is a continuous, high performance, intermediate modulus, PAN-based fiber provided in 12k filament count tows. Its density is 1.78 g/cm^3. The resin system, 8552, is a high-performance tough epoxy matrix for use in primary aerospace structures. This resin system is developed to operate in environments up to 250F. The density of 8552 is 1.3 g/cm^3.

For this specific manufacturing process, using the described materials, we are able to achieve a 40% fiber volume. We can now compute the composite density as

$$\rho_c = \rho_r \times f_r + \rho_m \times f_m = 1.78 \times 0.4 + 1.3 \times 0.6 = 1.492 \text{ g/cm}^3 \qquad \text{Eq. (8.9)}$$

The density of the resulting composite structure is 1.492 g/cm^3.

8.2.4 Modulus of Elasticity

Prior to discussing the computation of the modulus of elasticity, it is very important to highlight that, unlike density, the modulus of elasticity is orientation dependent. First, let us present the concept of **orthotropic** composite materials. Orthotropic composites are a subset of general anisotropic composites. The properties of orthotropic composites can be derived based on the individual properties along the orientation of the fiber and its perpendicular orientation. This is the case when we have two mutually perpendicular planes of symmetry in the material properties.

In these scenarios, the rules of mixtures can be applied along the orientation of the continuous fiber only. Other derivations – that we will not detail in this book – are needed to accurately account for the other orientations. One can imagine the required computations to be similar to the joining of two springs end to end or in parallel. The continuous/chain mode represents the final modulus of elasticity along the orientation E_c of the fiber, and the parallel arrangement represents the modulus of elasticity along the perpendicular orientation E_c^p.

Along the orientation of the fibers, we use the following computation:

$$E_c = E_r \times f_r + E_m \times f_m \qquad\qquad \text{Eq. (8.10)}$$

Along the perpendicular orientation of the fibers, we use the following computation:

$$E_c^p = \frac{E_r \times E_m}{E_m \times f_r + E_r \times f_m} \qquad\qquad \text{Eq. (8.11)}$$

8.3 Composites Manufacturing

Composites manufacturing deals with different manufacturing processes applied on composite materials to create parts and products. Individually, we can classify the different processes as additive, deformative, or assembly. Some composite manufacturing experts classify composites manufacturing as a subset of Additive Manufacturing, especially when dealing with the laminated object concept as classified and detailed in the Additive Manufacturing chapter. Others choose to classify some advanced concepts such as AFP as micro-welding processes.

Metals are isotropic since they exhibit the same physical properties in all directions. Composites are anisotropic in that the material exhibits different behaviors based depending on the orientation. Specifically, layered composites are, by their very nature, anisotropic since they have different properties in-plane compared to their thickness direction. Additionally, since fibrous composites typically have fibers in different layers in different directions, each layer has different properties, such as stiffness and strength. These characteristics mean that the anisotropic nature of composite parts must be carefully considered in their design and manufacturing.

The manufacturing of composite panels, due to the anisotropic properties of composite materials, required close attention to orientation. The orientation is more critical for parts with continuous fibers compared to those with short fibers and particles. There are multiple composite manufacturing techniques that are dedicated to specific structures. The development of these processes was aimed at achieving specific requirements for specific designs. In this book, we present eight different composite manufacturing techniques that are widely used. There are several additional techniques under development in advanced composites. Additional techniques will be added to subsequent editions of this book once they reached a certain stage of maturity and relevancy for industrial applications. In this section, we present the first seven techniques in individual sub-sections. And in the following section, we focus on AFP in greater detail as it is of particular interest to future airplane and aerospace engineers and is considered a "rising star" in composites manufacturing due to the distinct advantages this process promises.

8.3.1 Hand Layup

Hand layup is the manual stacking of composite layers according to the design. It is the most common and widespread manufacturing techniques for composites today. Individual layers are cut from woven or unidirectional sheets and are positioned in order to create the laminate. The typical steps in hand layup are (1) individual layer preparation, (2) bagging material preparation, (3) projection of boundary, (4) positioning of materials, (5) application of Resin System, (6) sealing, and (7) curing.

First we prepare all single layers. These layers are not necessarily of equal dimensions and can be of several different sizes. It is very common to include pad-up reinforcements between layers and at certain specific regions which results in a part of non-uniform thickness. Such built-up regions might be useful if holes are to be drilled or large cutouts are planned in late structural assembly. Note that the materials do not have to be similar. This step can include the preparation of honeycomb or other lightweight inserts as well. These would act as a core in a sandwich structure if placed in specific locations to reduce the weight of the overall part or product.

The next step is the preparation of the bagging material and placement on the tool surface. We include a first layer of the bagging material, such as a nylon bagging film, that is designed to be used in the curing cycle which could be in an autoclave or oven/RTM process. It is very important to understand the operating conditions of the bagging material to avoid burning or otherwise damaging it in the next steps.

The third step is not required for all industries and applications. However, it is often desired to have support from a projection system that shows the exact location of each individual layer to increase precision and repeatability for selected critical applications. The projection of layer boundaries on the table or mold facilitates the reduction of human error and can support a more time-efficient production process. For some parts and products that require certification, the use of a projection system can be a factor supporting fulfillment of the certification requirements since it leads to more consistent parts and repeatable processes.

The fourth step is the position of the fabric. If a projection system is used (see step 3), then the operator places the single layer prepared in step 1 within the boundary provided. If no projection system is available, the technician must place the single layer based on the available documentation, often in form of drawings or layup plans (paper or digital).

Following, in step 5, the resin system will be applied by the operator. She/he often uses a hand roller to ensure equal repartition of the resin. Steps 4 and 5 are continuously repeated as needed to build the laminate layer by layer. Each layer must be placed precisely on the layer below, including all fabric, adhesives, and cores, etc. As this involves manual operations, both steps 4 and 5 can lead to significant deviations of quality depending on the operator's performance. This is one of the major critiques of the manual layup technique and key motivation to develop more automated, and thus controlled and repeatable, composite manufacturing processes.

Finally, we seal the bagging process in the sixth step, and proceed with the curing of the composite in a final step 7, typically in an autoclave or oven. The use of a peel ply is important to remove the part from the tool.

8.3.2 Filament Winding

This manufacturing process is particularly suitable for pressure vessels and structures with circular cross sections. The manufacturing process is rather simple: the structure is placed on a rotating mandrel that simultaneously pulls the fibers and places them. As we learned, simplicity is a good starting point for any manufacturing process as long as it leads to the desired outcome – in this case, composite parts with uniform cross-sectional area. These continuous filaments (reinforcement material) are guided from a creel onto the structure. In between, they are impregnated with the resin (matrix material) by being dipped into a resin bath in situ (in process). The creel holds the fibers in their dry format, without any wetting or with only limited toughening agents added. The separators ensure uniform impregnation and the final guiding ring merges the fibers prior to being placed on the structure as shown in Figure 8.8.

There are multiple proven layup strategies and patterns for filament winding available. One of the most common ones is labeled as hoop winding, where the filament is placed circumferentially at almost 90° to the axis of the part. Also, particular attention is needed at the tips of the structure (end closures). For structures that are designed to operate under certain pressure conditions, we can expect the end closures to be mechanically fastened to the filament wound structure.

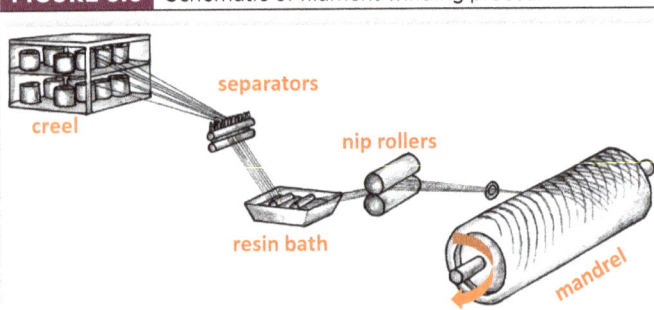

FIGURE 8.8 Schematic of filament winding process.

© Samar Mouawad

8.3.3 **Pultrusion**

Pultrusion attempts to join several steps into one manufacturing process. This inherently creates a manufacturing quality inspection issue, as it is difficult to pinpoint the root cause and at which step a problem arose.

The two main steps that pultrusion combines are the impregnation of fibers as they are withdrawn from racks and routed through guides, and the collimation of fibers into aligned bundles prior to entering the die which provides the profile/shape.

The term pultrusion is a hybrid between pulling and extrusion. The fibers are pulled in a manner similar to wire and bar drawing, i.e., from the exit point. Hoa (2018) states that the state of development of pultrusion as a process for composite manufacturing is still in the experimental stage. This status is due to the integration of many inherently complex individual steps into the process that makes it complex.

8.3.4 **Composite Molding**

Composite molding is similar to die casting processes, in a sense that a composite preform is pressed along a mold and heated or impregnated to initiate a curing process. Following this step, we de-mold the part and retrieve it for further finishing processes such as drilling and trimming.

8.3.5 **Automated Tape Laying (ATL)**

Automated Tape Laying (ATL) is a versatile manufacturing process that is shape independent. It provides access to create different shapes using computer guided numerical controls. The starting material of ATL is unidirectional tapes. Width of those tapes is usually 3 in. and above. While excellent for productivity, the width will prevent manufacturing of complex shapes with different curvatures. This was the precursor to the creation of Automated Fiber Placement (AFP), the most widespread and versatile composites manufacturing technique. AFP is covered in detail in the following section, as it represents a cornerstone of composites manufacturing education.

8.4 **Automated Fiber Placement (AFP)**

AFP is increasingly seen as the premier manufacturing process for high-quality composite structures. AFP provides numerous advantages over alternative composites manufacturing processes. While similar in nature to ATL, in the sense that a machine is accurately placing the composite material using computer numerical control machines, AFP places tows (<=1 inch width) while ATL places tapes (3=< inch width). This reduced width of the to-be-placed material offers higher flexibility in terms of possible designs. Which in turn can result in more complex structures with regard to both geometry and tailorability. Indeed, as we will demonstrate in the subsequent paragraphs, AFP machines offer the capacity to individually place and cut tows with widths of less than 1 inch. Moreover, AFP is gaining momentum and industrial adoption is increasing rapidly as knowledge of this still relatively new process is developing fast and the scientific community is growing steadily. Most aerospace companies are now equipped with AFP machinery or at the very least have access to AFP production resources. Finally, AFP also generates less waste material than other composite processes and is a highly repeatable process. Given the price point of high-quality composite raw materials, this reduction of waste is not only beneficial for the environment but also increases the economic feasibility as well as facilitates the broad adoption of advanced composites manufacturing.

AFP machines are generally gantry based or robotic based. Both systems include a computer-controlled placement head which moves to precisely place material. Gantry-based machines have the placement head mounted on a gantry-like structure, usually offering a large building envelope. Robotic-based AFP machines have the placement head mounted as the end effector of a robotic platform. The robot's maneuverability and reach defines the building envelop. The end effector includes all needed components to achieve process control during the laminate manufacturing. Process parameters that are tightly controlled include temperature, speed, and pressure, etc. (Figure 8.9).

AFP machines are a system of several distinct components, such as the creel, tool, mandrel, and placement head. In the following section we present a brief introduction to each of the essential parts of an AFP system.

Creel: The main function of the creel is to preserve the material at the right temperature pending usage. It typically holds the material spools and has a separator to dissociate the tow protecting film from the tow itself. Creels are typically used for gantry-based systems, but some robotic-based systems do also utilize creels. The number of spools held by the creel is typically limited by how many tows the AFP head can deposit at the same time.

Head: The AFP head contains the elements that predominantly control the process parameters such as the roller, compaction, tensioner, cutting mechanism, and heating mechanisms. AFP heads can be rigid (robotic based) or can include mechanisms to facilitate tow delivery increase machine accessibility and visibility.

Roller: The AFP roller represents the contact point between the machine and the underlying laminate (future part or product). Roller material and dimensions are typically set by the corresponding composite material. Rollers also absorb some of the compression as the head moves towards the tool for placement.

Heating Mechanism: One of the most important elements in the placement process is to achieve uniform heating of the substrate. This uniform heading leads to good and consistent adhesion. Multiple heating sources and methods have been investigated during the last few decades, including infrared (IR), nitrogen gas, laser, and a combination. Although IR heaters are the predominant heating mechanism today, recent research efforts to enhance AFP manufacturability for thermoplastic materials are offering new, more appropriate heating mechanisms to meet thermoplastics requirements.

Mandrel: The mandrel holds the tool for the layup process. It is very important to highlight that in AFP terminology we talk about tool in reference to where the material will be laid. The term "mold" (see Chapter 2 "Deformative Manufacturing") is typically not used in the AFP community. The mandrel can be flat, to hold planar and complex tools, or it can be a cylindrical structure to enable designs such as an airplane fuselage.

FIGURE 8.9 Robotic-based AFP machine: NASA's ISAAC (Harik, 2019).

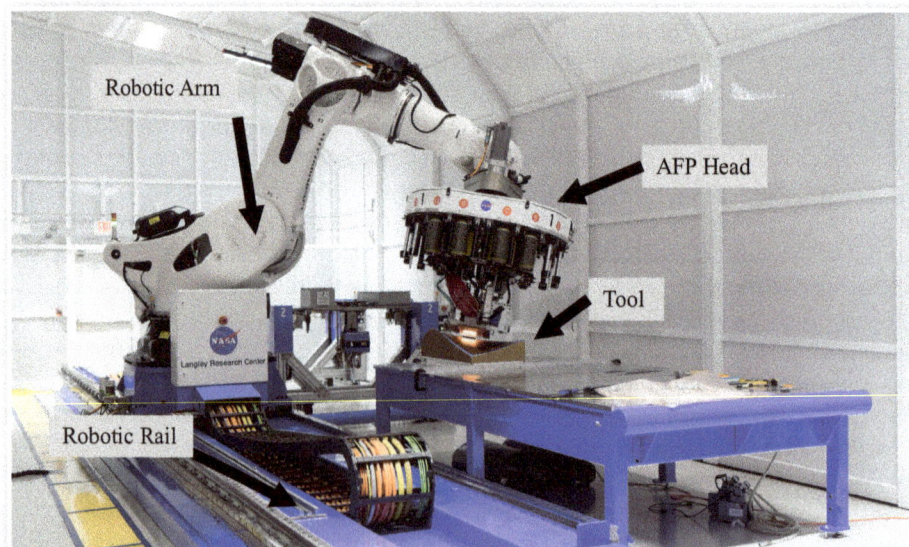

FIGURE 8.10 AFP process terminology: Laminate, Ply, and Course.

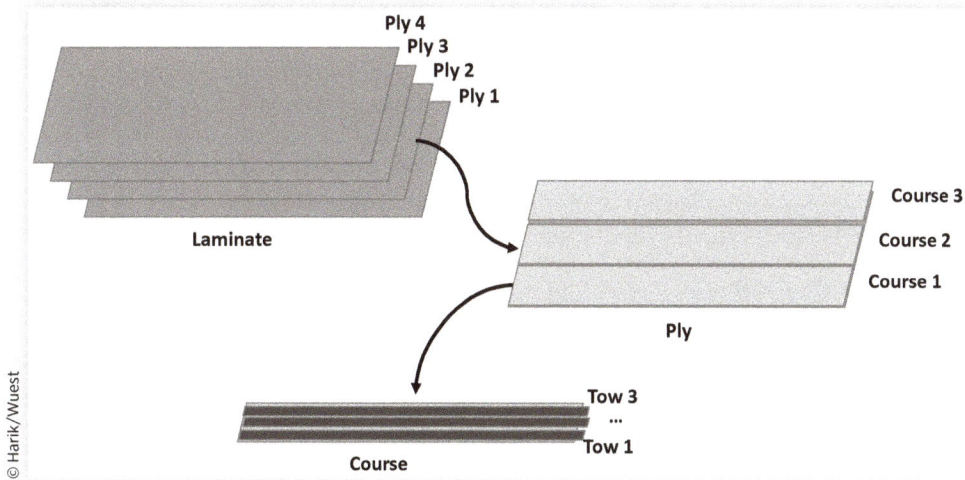

© Harik/Wuest

8.4.1 **AFP Process**

The AFP process offers a high degree of flexibility, as the laminate is built by aggregation of slit tapes (tows) with very small width, such as ½ or 1 in. The aggregation of the individual tows into one placement (one motion of the machine) identifies a course. A succession of multiple courses creates a single ply. The placement of the different plies creates the laminate (Figure 8.10).

AFP includes four basic components:

1. *Feeding mechanism*: Ensures both tension and a proper feed adjustment of the tows

2. *Cutting mechanism*: Offers the ability to cut individual tows providing extreme flexibility in the context of laminate building

3. *Heating mechanism*: Heats the underlying layer prior to placement of the new one

4. *Pressure mechanism*: Applies pressure through the roller to squeeze out any voids and ensure tackiness/consolidation (attachment) of the new layer onto the previous one

The basic processes of AFP are continuously under investigation. There is no single method to provide any of the four elements needed for the AFP process. For example, the heating mechanism can be provided by hot gas, laser, or IR heaters. Xia et al. (2018) present a solution to the AFP transient radiative heat transfer problem to achieve an ideal nip point. The need to achieve uniformity over complex surfaces and the context of closed-loop/open-loop are detailed. IR heating has grown in popularity throughout the aerospace industry as the preferred method of producing ideal material properties for AFP processes. It offers positive attributes such as better efficiency in open areas and less power consumption (Figure 8.11).

8.4.2 **AFP Defects**

While AFP offers many benefits, there are manufacturing defects which can occur. These defects can reduce the final part's ability to support load so they must be either corrected prior to the placement of the subsequent layer. The predominant AFP defects are listed below, as presented in Harik et al. (2018) (SAMPE).

Gap/Overlap: A gap emerges when two adjacent tows are not perfectly placed next to each other, resulting in an unintended space between the two tows. An overlap happens when one tow overlaps the adjacent tow. The most common cause of gaps and overlaps is

FIGURE 8.11 Schematic of AFP Process.

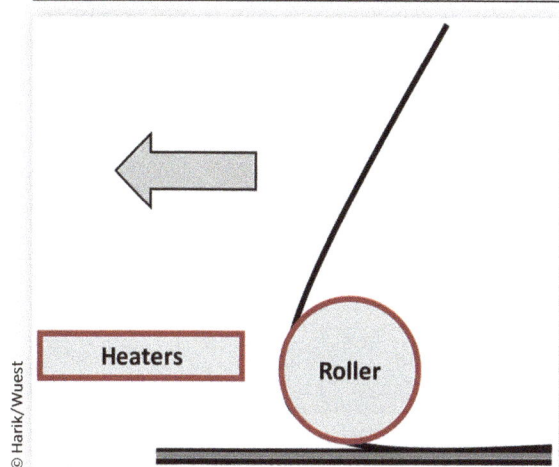

© Harik/Wuest

steering during layup. However, gaps and overlaps occur outside of steering when laying up over a complex 3D-shaped tool surface (Figure 8.12).

Pucker: A pucker typically initiates at the inside radius of a steered tow and results in the tow lifting from the tool surface. The pucker could occur either partially or across the entire tow width. The tow forms an arch of excess material that does not adhere to the underlying substrate material. Puckered tows are caused by the excess feeding of a tow that gradually accumulates ahead of the compaction roller and at some point emerges in the part surface. If placement is over a compliant surface, with the force of the compaction roller, longer tows may be deposited that can form the pucker after the surface springs backs to its original shape (Figure 8.13).

Wrinkle: A wrinkle is typically indicated by a wavy pattern of puckering along the edge of a tow when it is steered through a non-geodesic path over a complex (potentially doubly-curved) surface or following a steered path on a flat surface. These types of defects occur on the inner radius and remain out of plane after compaction and curing. Wrinkles are often caused by placing tows at small steering radii, which can lead to excessive differential length between the two edges of the projection of the tow on the part surface. Since the two edges of a tow delivered from the machine head are equal length, part of the excessive differential length presents as puckers and/or wrinkles (Figure 8.14).

Bridging: A bridged tow does not fully adhere to the concave surface (female tool portion) or a re-entrant corner or ramp-up area over which the tows are being laid up on. This results in a gap between the radius of the concave tool surface and the tow. The main causes of a bridged tow are too much tension on the tow, which will force the tow to lift up, or insufficient tack adhesion to the surface the tow is laid up on due to the roller not providing full contact with the substrate material (Figure 8.15).

Angle Deviation: Angle deviation is when the angle of the as-manufactured layup deviates from the as-designed one. Angle deviation can be caused by incorrect roller coverage or small radius steering as the tow may move after being steered (Figure 8.16).

Fold: This defect occurs when the tow folds in the transverse direction onto itself, creating a gap in the surface coverage and doubling the tow thickness over the folded part. An extension (and probably the worst-case scenario) of the folding could be rolling (or completely twisting) of the tow to become "rope" like. Tensioner errors, such as either insufficient tension or too much tension, could increase the probability of the tow to fold. Long unsupported/complex tow paths from the spools to the head can also result in folding. In a steered/curved tow path, the outer segment of the tow may fold towards the inner side after the compaction roller nip point due to tension on the outer edge of the tow and improper tack adhesion (Figure 8.17).

FIGURE 8.12 Gap/overlap CAD representation.

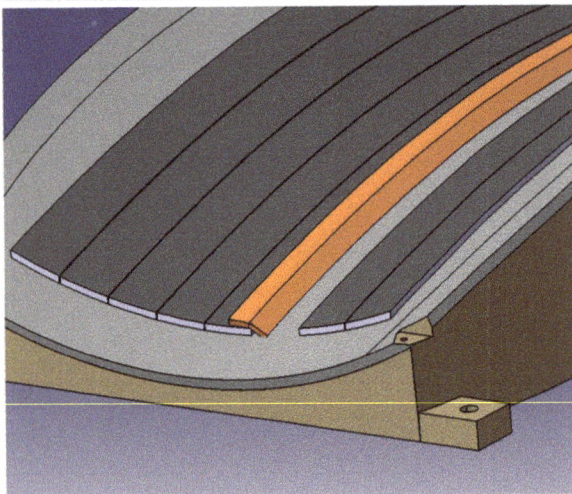

© R. Harik, C. Saidy, S. Williams, Z. Gurdal, B. Grimsley, *SAMPE 2018*

FIGURE 8.13 Pucker CAD representation.

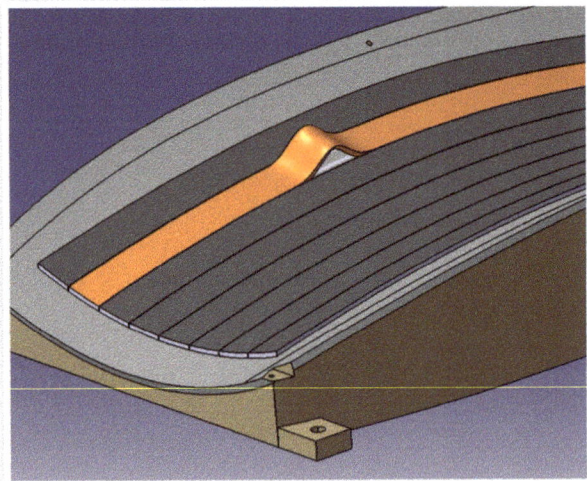

© R. Harik, C. Saidy, S. Williams, Z. Gurdal, B. Grimsley, *SAMPE 2018*

FIGURE 8.14 Wrinkle CAD representation.

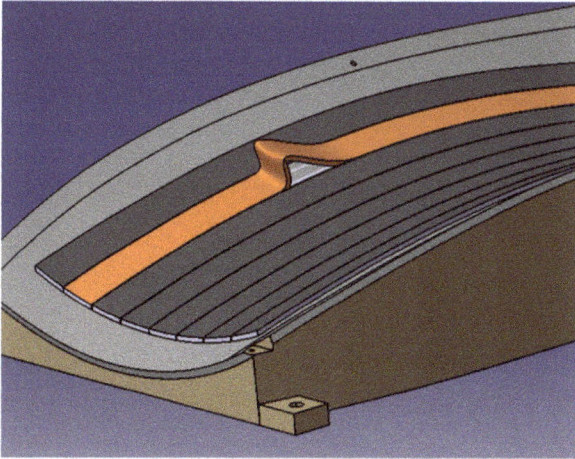

© R. Harik, C. Saidy, S. Williams, Z. Gurdal, B. Grimsley, *SAMPE 2018*

FIGURE 8.15 Bridging CAD representation.

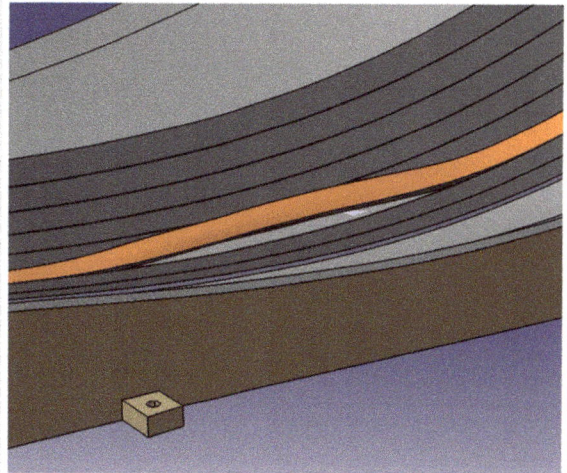

© R. Harik, C. Saidy, S. Williams, Z. Gurdal, B. Grimsley, *SAMPE 2018*

FIGURE 8.16 Angle deviation CAD representation.

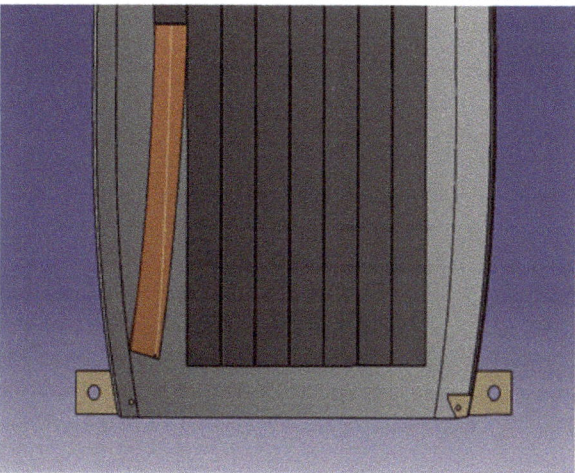

© R. Harik, C. Saidy, S. Williams, Z. Gurdal, B. Grimsley, *SAMPE 2018*

FIGURE 8.17 Fold CAD representation.

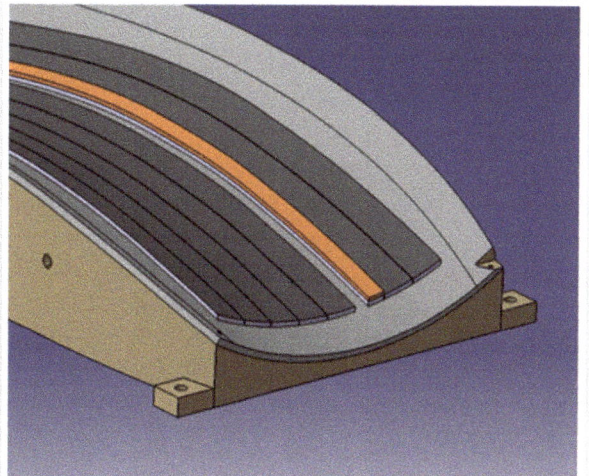

© R. Harik, C. Saidy, S. Williams, Z. Gurdal, B. Grimsley, *SAMPE 2018*

CHAPTER 8

Twist: A twisted tow is where the tow rolls axially 180° onto itself and is then flattened by the compaction roller. Depending on the length over which the twisting occurs, the shape may be like a bow tie with bunching of the fibers and increased thickness at the center. Twisted tow could be initiated by folding, in which the fold grows and completes a full turn rather than unfold (folded tow could be considered incomplete twist). Friction between guide holes along a long/complex tow path and a tacky tow may cause twisting due to head rotation during bidirectional layups (Figure 8.18).

Wandering Tow: A wandering tow is when the portion of the tow between the roller and the cutter wanders from the original fiber path after being cut. Similar to tow "angle deviation," wandering tows are more attributable to having an unsupported portion of the tow between the compaction roller and the tow cutter, and therefore the angle deviation will only be of the dimension of this unsupported tow length (Figure 8.19).

Loose Tow: A loose tow generally refers to a section of a tow (or tows) that the machine head attempts to place on a part without having complete and precise control over where it is actually placed, causing the tow to meander. A tow is completely loose when the length of a tow

FIGURE 8.18 Twist CAD representation.

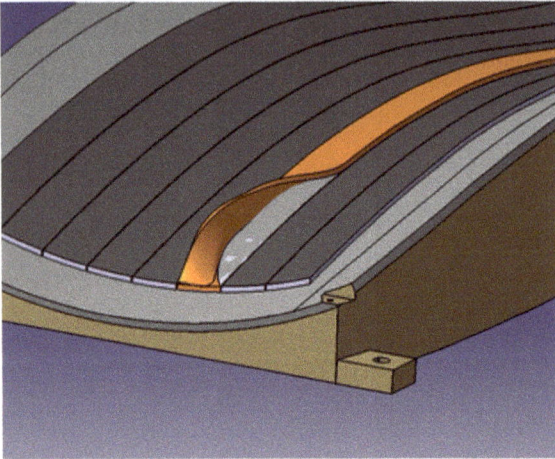

FIGURE 8.19 Wandering tow CAD representation.

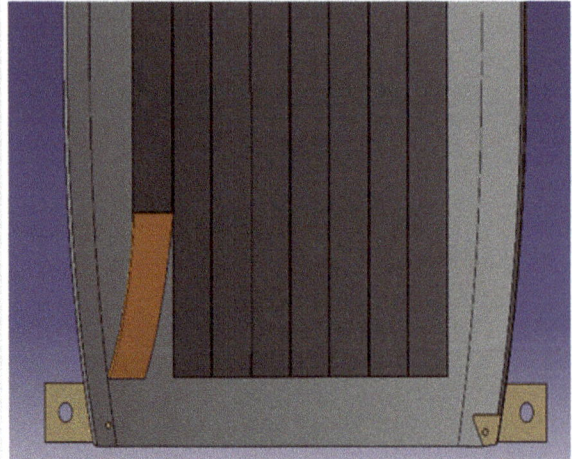

is shorter than the distance between the cutters and the compaction roller that controls tow's final position. In this case, the tow is free to land on an arbitrary position. If at the end of a course the fiber path is still steered, the section of the tow before the compaction roller may not follow the defined steered path.

Splice: A splice is created when two tows are joined together by the material or slitting supplier end to end in a spool by overlapping the tape in preparation for tacking together and slitting. Theoretically, carbon fibers can be drawn infinitely long. However, most AFP pre-impregnated tows are slit-tapes that are cut from a roll of finite-length unidirectional tape. These slit tapes are spliced and spooled based on customer specifications. Typically 1 to 3 in of the two tapes are connected, resulting in a portion of the tow that is thicker than the rest and is usually marked by white dashes for easy detection (Figures 8.20).

Foreign Object Detection: A foreign object debris (FOD) defect is when a small piece of composite material, either carbon fiber "fuzz ball" or "resin ball" that has collected on surfaces of the head or other debris from the production area fall onto the part during layup. More generally, FOD includes any debris that ends up in the part. This results in a small excess volume of material on the ply if laid up over (Figures 8.21).

FIGURE 8.20 Splice CAD representation.

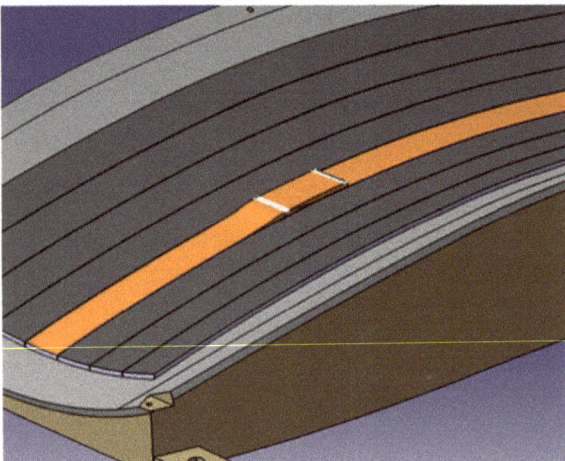

FIGURE 8.21 FOD CAD representation.

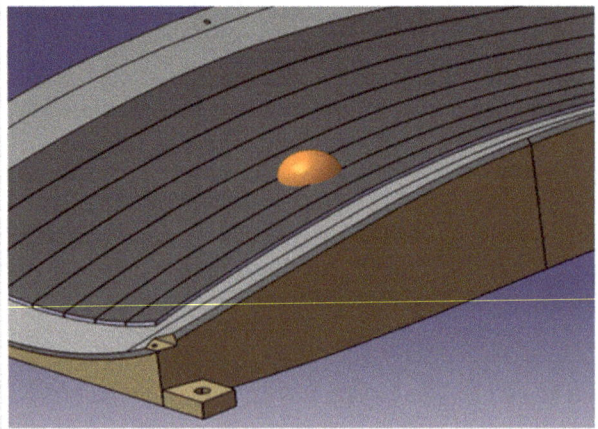

8.4.3 AFP Layup Strategies

This section is adapted from Rousseau (2019). Manufacturing an optimal composite part requires the selection of an optimal layup strategy similar to the concept of toolpath generation in subtractive manufacturing, as mentioned in Chapter 6. Layup strategies are usually a 'forced' element, as we typically want to obtain constant angle laminates. However, this becomes impossible based on the different geometries. Therefore, we recourse to other methods that can provide optimal coverage while minimizing the AFP defects as presented in the previous section. These methods are often split into constant angle, constant curvature, geodesic, and other. They are additionally complemented with a propagation strategy for complete coverage. Table 8.3 compares the features of the most prominent layup strategies. For further details refer to Rousseau (2019).

8.4.4 AFP Process Planning

This section is based on Halbritter (2019) and represents the matchmaking process between design and manufacturing: Process planning. The study conducted by Halbritter et al. (2019) represents a survey of most important process planning functions. Figure 8.22 illustrates the most important functions and the time taken through the process. Table 8.4 represents the most important functions, with their expected output and additional remarks.

8.4.5 AFP Inspection

This section is based on Sacco et al. (2019) and represents the inspection of the AFP quality. Figure 8.23 presents the developed tool that recognizes manufacturing imperfections (such as AFP defects) and their recognition/classification based on machine learning algorithms. AFP inspection is very important as it is currently a manual process with several efforts to created automated tool and software.

TABLE 8.3 Overview of different AFP layup strategies

Geometry of the surface					
Strategy for the reference curve	Coverage strategy	Coverage percentage		Notable defects	Ranking
		% of gaps	% of overlaps		
Geodesic path	Fast marching method				
	Parametrical parallel curves				
	Shifted curves				
	Independent curves				
Constant Curvature	Fast marching method				
	Parametrical parallel curves				
	Shifted curves				
	Independent curves				
Linear variation	Fast marching method				
	Parametrical parallel curves				
	Shifted curves				
	Independent curves				
Use of control points	Fast marching method				
	Parametrical parallel curves				
	Shifted curves				
	Independent curves				
Following the constraints	Fast Marching Method				
	Parametrical parallel curves				
	Shifted curves				
	Independent curves				

FIGURE 8.22   Time consumed by different process planning functions.

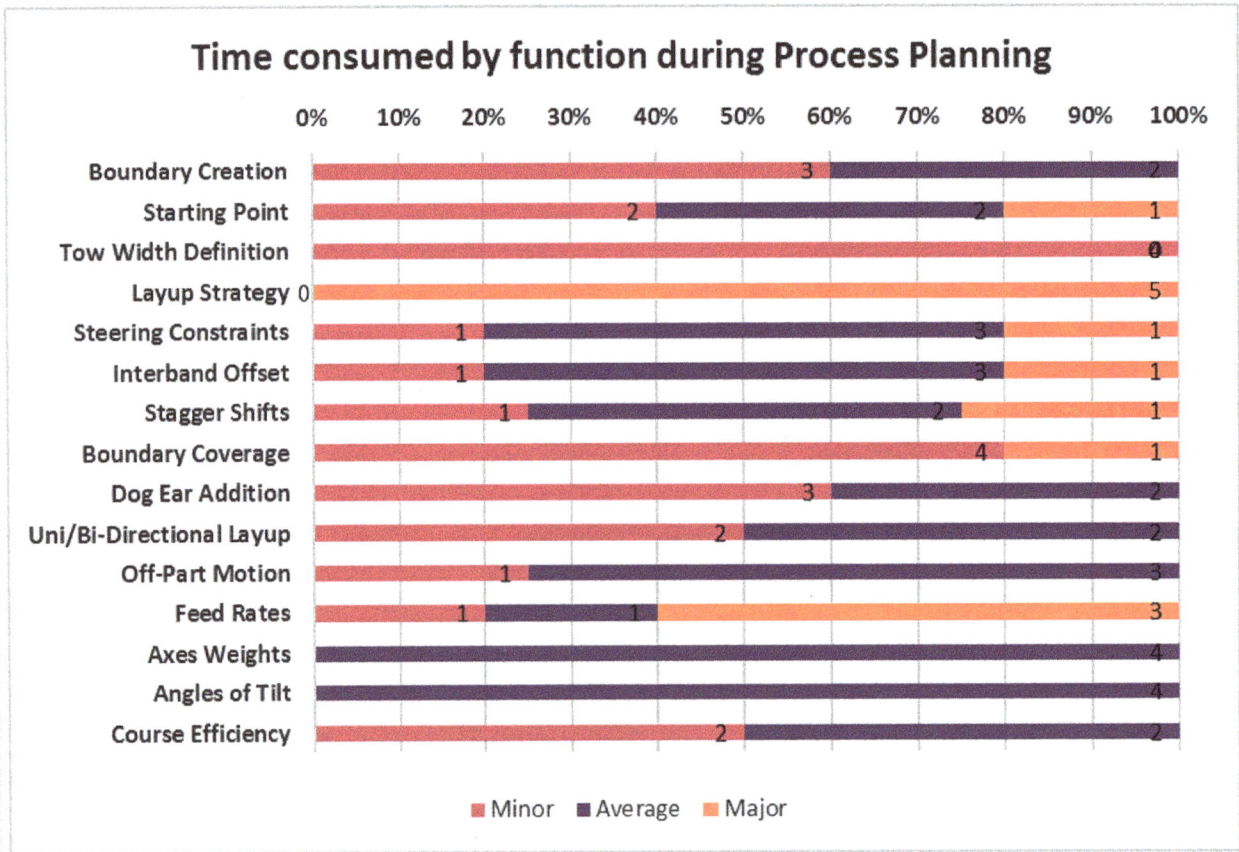

Time consumed by function during Process Planning

TABLE 8.4 Principal process planning functions

#	Function title	Expected output	Additional remarks
4	Layup Strategies	• Provide a set of metrics for each layup strategy relevant to defect likelihood • Optimize based on user priorities such as angle deviation, steering, gaps/laps, etc.	• Determined by engineering with some input from numerical control programming • Choose the strategy that minimizes through-thickness deviation
9	Dog Ears	• Modify ply boundaries to account for additional material placement • Determine optimal dog ear strategy by structural analysis	• Accept default dog ear method defined by process planning if they are not engineering dependent
7	Stagger Shift	• Reduce coincident laps/gaps through laminate thickness through modification of the starting point	• Consider the importance of reducing the feature that stagger shift is attempting to minimize • Combine stagger shift functionality with the start point function
5	Steering Constraints	• Dynamically vary the acceptable steering radius over the surface depending on local curvature • Recommend minimum steering radius based on geometry and layup orientation	• Only suggest minimum steering radius as guideline, since the surface curvature will alter the effects of in-plane curvature
2	Starting Point	• Minimize steering error, gaps, overlaps, and angle deviation, in addition to meeting course stagger requirements by placement of the starting point	• Finding proper placement varies by path generation algorithm, and can be a tedious, but process critical action

FIGURE 8.23 Defect classification and detection based on machine learning.

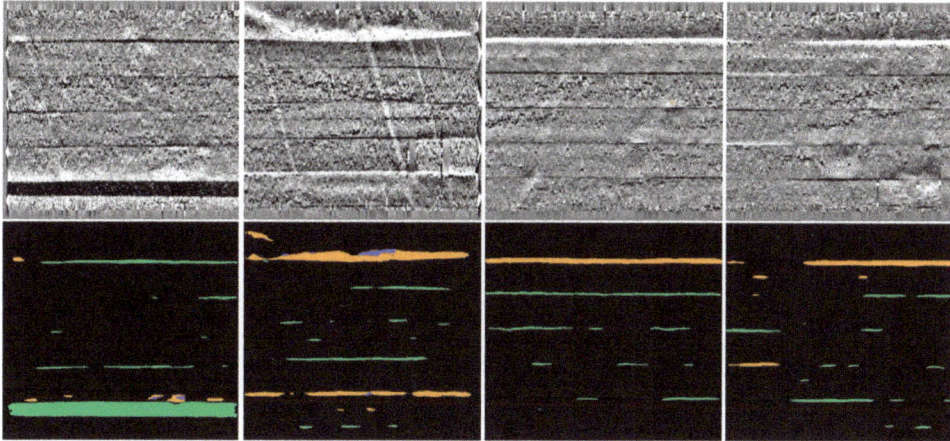

© C. Sacco, A. Baz Radwan, T. Beatty, and R. Harik, *SAMPE 2019*

8.4.6 **AFP Case Study**

This section is based on (Harik, 2019) and represents a case study of using AFP in the context of wind tunnel fabrication. (Harik, 2019) also includes procedure of how to manufacture a part using AFP, including: Design Assessment, Computer Aided Process Planning, Tool Form Manufacturing, Dry Run, and placement.

CAPP results in a functional NC Code, whereas the Tool Form Manufacturing results in a functional tool to be used in the process. Before we initiate the manufacturing, a step of dry run, where we simulate the machine motion on the tool without material (see Figure 8.24).

Figure 8.25 demonstrates the manufacturing and inspection of the part, references in (Harik, 2019). The figure show the heater in action as the prepreg is being laid on the tool. Figure 8.26 demonstrates the final part.

FIGURE 8.24 Manufacturing flowchart.

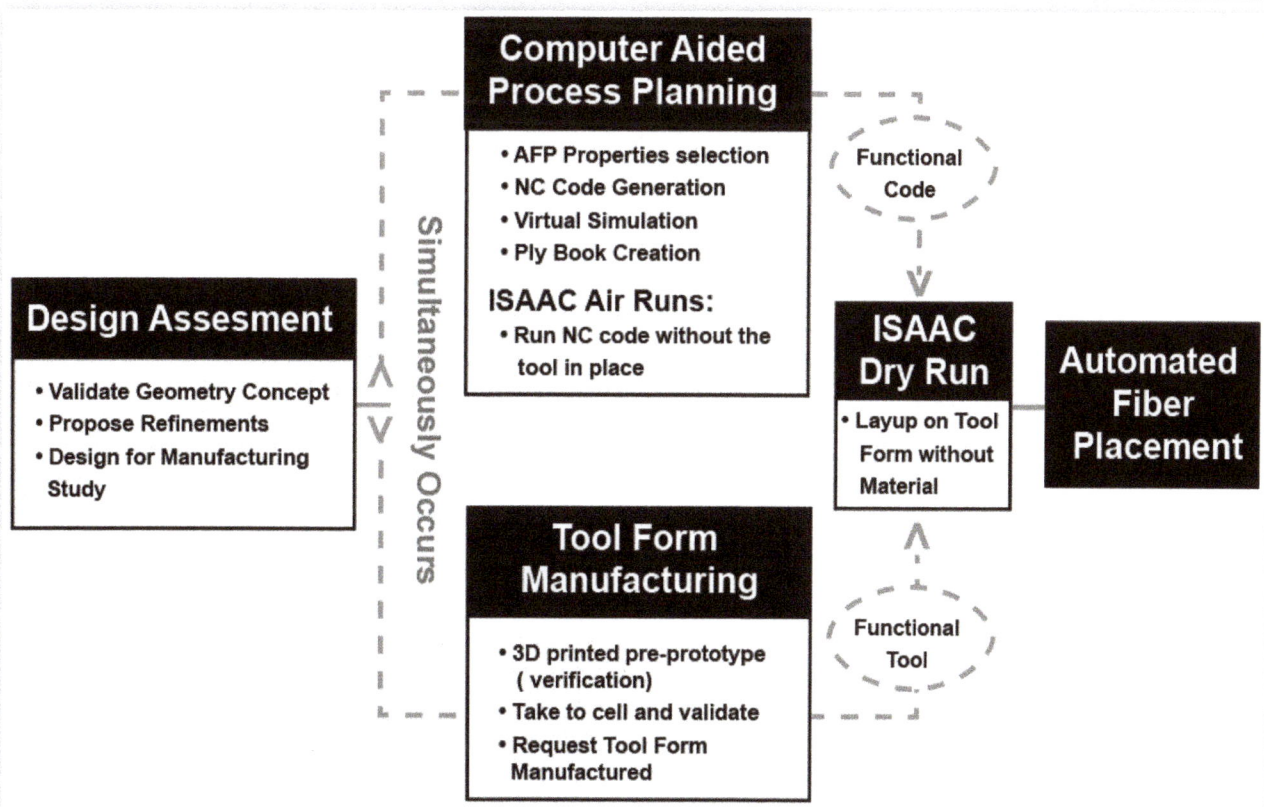

© R. Harik, J. Halbritter, D. Jegley, R. Grenoble, *SAMPE 2019*

FIGURE 8.25 (a) Layup process. (b) Inspection, discussion and correction; layup shows splices (defect) whose tows will be replaced.

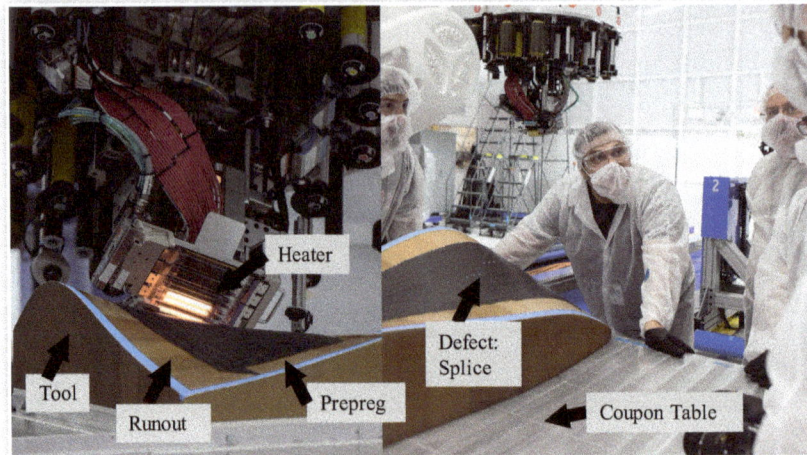

© R. Harik, J. Halbritter, D. Jegley, R. Grenoble, *SAMPE 2019*

FIGURE 8.26 Manufactured part.

© R. Harik, J. Halbritter, D. Jegley, R. Grenoble, *SAMPE 2019*

References

Ahmed, H., van Tooren, M., Justice, J., Harik, R. et al., "Investigation and Development of Friction Stir Welding Process for Unreinforced Polyphenylene Sulfide and Reinforced Polyetheretherketone," *Journal of Thermoplastic Composite Materials* (2019), Accepted for publication.

Albazzan, M., Harik, R., Tatting, B., and Gurdal, Z., "Efficient Design Optimization of Nonconventional Laminated Composites Using Lamination Parameters: A State of the Art," Composites Structures, Accepted for publication, 2019.

Bahamonde, L., AlBazzan, M., Chevalier, P., Gurdal, Z. et al., "Rapid Tools for an AFP Manufacturing Defects Assessment Framework," *Submission to SAMPE 2018 Conference & Exhibition*, 21 – 24 May 2018, Long Beach, California, US, 2018.

Halbritter, J., Harik, R., Saidy, C., Noevere, A. et al., "Automation of AFP Process Planning Functions: Importance and Ranking," *SAMPE 2019 Conference & Exhibition*, Charlotte, North Carolina, US, May 2019, 20–23.

Halbritter, J., Tuk, A., Harik, R., van Tooren, M. et al., "Modular Conversion Tool for Generating Fiber Placement Code from Optimized Fiber Steering Path," *The Composites and Advanced Materials Expo 2017 (CAMX 2017)*, September 11-14, Florida, United States, 2017.

Harik, R., Halbritter, J., Jegley, D., Grenoble, R., "Automated Fiber Placement of Composite Wind Tunnel Blades: Process Planning and Manufacturing," *SAMPE 2019 Conference & Exhibition*, Charlotte, North Carolina, US, May 2019, 20–23.

Harik, R., Saidy, C., Williams, S., Gurdal, Z. et al., "Automated Fiber Placement Defect Identity Cards: Cause, Anticipation, Existence, Significance, and Progression," *Submission to SAMPE 2018 Conference & Exhibition*, 21 – 24 May 2018, Long Beach, California, US, 2018.

Hoa, S., *Principles of the Manufacturing of Composite Materials* (Lancaster, PA: DEStech Publications, Inc., 2018).

Rousseau, G., Wehbe, R., Halbritter, J., and Harik, R., "Automated Fiber Placement Path Planning: A State-of-the-Art Review," *Computer Aided Design and Applications* 16, no. 2 (2019): 172-203, doi:10.14733/cadaps.2019.172-203.

Sabido, A., Bahamonde, L., Harik, R., and vanTooren, M.J.L., "Maturity Assessment of the Laminate Variable Stiffness Design Process," *Composite Structures* 160 (2017): 804-812.

Sacco, C., Baz Radwan, A., Beatty, T., and Harik, R., "Machine Learning Based AFP Inspection: A Tool for Characterization and Integration," *SAMPE 2019 Conference & Exhibition*, Charlotte, North Carolina, US, May 2019, 20–23.

Sacco, C., Baz Radwan, A., Harik, R., and van Tooren, M., "Automated Fiber Placement Defects: Automated Inspection and Characterization," *Submission to SAMPE 2018 Conference & Exhibition*, 21 – 24 May 2018, Long Beach, California, US, 2018.

Sarikaya, I., Tahiyat, M., Harik, R., Farouk, T. et al., "Plasma Surface Functionalization of AFP Manufactured Composites for Improved Adhesive Bond Performance," *SAMPE 2019 Conference & Exhibition*, Charlotte, North Carolina, US, May 2019, 20–23.

Wehbe, R., Tatting, B., Harik, R., Gurdal, Z. et al., "Tow-Path Based Modeling of Wrinkling During the Automated Fiber Placement Process," *The Composites and Advanced Materials Expo 2017 (CAMX 2017)*, September 11-14, Florida, United States, 2017.

Xia, K., Harik, R., Herrera, J., Patel, J. et al., "Numerical Simulation of AFP Nip Point Temperature Prediction for Complex Geometries," *Submission to SAMPE 2018 Conference & Exhibition*, 21 – 24 May 2018, Long Beach, California, US, 2018.

Zhang, Y., De Backer, W., Harik, R., and Bernard, A., "Build Orientation Determination for Multi-material Deposition Additive Manufacturing with Continuous Fibers," *Procedia CIRP* 50 (2016): 414-419, ISSN 2212-8271.

CHAPTER 8

Manufacturing Quality Control and Productivity

The previous chapters of this book focused on different processes, tools, and techniques of transforming a part and adding value. In this chapter, we will focus on how to ensure that the selected process will achieve the intended outcome. The desired outcome of a manufacturing process can generally be described as fulfilling the design requirements, such as tolerances, dimensional accuracy, material properties, and surface quality, thus achieving the desired quality objective (Figure 9.1). Variance in the intended output is defined as variability, which is most commonly measured through range and standard deviation. Quality and variability are inversely proportional – with the quality goal being as consistent and predictable as possible.

Today, manufacturing quality is not only a distant or visionary goal but a necessity in order to stay competitive on the marketplace. Of all the "-ilities," such as production variability or product reliability, Quality is the most dominant concern in the literature on engineering systems, even more widely discussed than safety and reliability (De Weck et al., 2012). For example, in the automotive industry, suppliers are often contractually obligated to guarantee "zero defect" parts to be delivered to the car company for final assembly. This guarantee has to be backed up by evidence and data that helps the OEM to evaluate the suppliers' ability to achieve this "zero defect" goal. In case they nevertheless fail to deliver the required quality, their contracts can be canceled and they can be excluded from future bids for a certain period of time.

There is a variety of tools, standards, and methods available that are intended to help manufacturing companies managing their process and product quality. On one side of the spectrum are technological innovations such as vision systems that allow for automated in situ quality inspection of, e.g., surface quality. On the other side, we have statistical tools that help to approximate the overall quality of the process and products. In between there is a wide variation of quality-centered approaches, including preventative tools as well as control algorithms to adjust parameters in real time to achieve the desired quality outcome. In this chapter, we focus on established quality tools, standards, and methods, while we do not emphasize on specific technological innovations in detail. There are simply too many technological innovations available today and we regard them as useful technological support within the overall quality strategy of a manufacturing firm. For example, in a Failure Mode and Effects Analysis (FMEA) (see Section 9.2.3), a high-resolution light-field vision system might be a solution for an identified critical failure mode. However, the camera by itself serves little purpose when the quality strategy is not developed accordingly.

FIGURE 9.1   Quality as a main objective of manufacturing companies.

© NicoElNino/Shutterstock

9.1 **Quality Definition**

Quality is defined as "the degree to which a set of inherent characteristics fulfils requirements" with requirements understood as the "need or expectation that is stated, generally implied or obligatory" (DIN EN ISO 9001:2008). A product can inherit different qualities, with the sum of a set of quality instances, such as security, workmanship, or durability cumulatively representing the final product quality (Kamiske and Brauer, 2008; Wuest, 2015). In a manufacturing environment, there are different views and definitions available, ranging from very technical perspectives to rather generic. An example of a more general view is postulated by Garvin (1984) that states that manufacturing quality is basically "confirming with requirements." A more technical definition is the view that manufacturing quality is "primarily a factor of machining tolerances" (Kaiser, 1998). Increasingly the concept of Perceived Quality (PQ) is relevant to attract customers. PQ focuses on the perception of a products quality by the users relative to other products. However, this is mainly related to design, while during manufacturing the manufacturing quality is in the focus.

9.1.1 **Product Quality**

The final outcome of a manufacturing process can be either a final product, a part, or an intermediate product. However, in order to reduce the complexity, in this chapter we assume that the final output of the process is the "final product" (or product) with distinct requirements according to which we can measure its product quality. In this case, the term "product" includes also parts, components, and workpieces. Figure 9.2 illustrates how product quality is measured by the fulfillment of the (quality) requirements given the characteristics of the final product. Product quality is determined by the process quality, which in turn is influenced by the commands

FIGURE 9.2   Elements of quality (adapted from Masing, 2007; Sitek, 2012; Wuest, 2015).

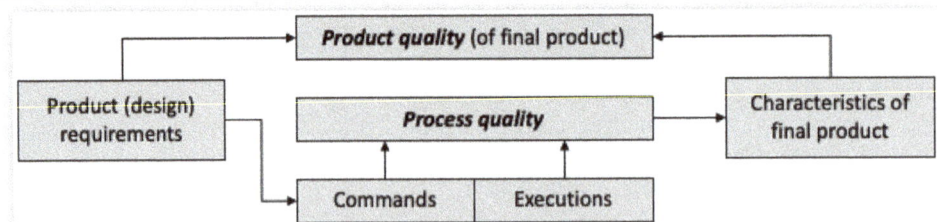

© Wuest, 2015, Masing, 2007, Sitek, 2012

and executions derived from the design requirements (Wuest, 2015). Each manufacturing process within the value chain leads to a product with a certain quality. However, we also have to keep the whole manufacturing chain in mind when planning for the final product quality. The reason is that while at each stage the quality requirement might be fulfilled, there are interrelations across processes that are not understood and included in the quality judgment yet have an impact on the final quality outcome. An example for this phenomenon is tolerance stackups and distortion caused by activated residual stress.

Some quality characteristics can be measured more easily than others, such as length, depth, or weight, commonly summarized as dimensions/metrology. Other characteristics are more difficult to measure, such as functions or aesthetics. Dimensions have the advantage of being easier to monitor and control. In manufacturing quality, the characteristics are mostly physical in nature, making them easier to measure and quantify compared to softer characteristics such as smell or aesthetics. Nevertheless, there are some quality characteristics that are only measurable by proxy or in a destructive manner such as residual stress. This makes it more complicated to judge the quality of a part holistically and emphasizes the importance of statistical methods to reduce the need for a 100% in situ monitoring.

9.1.2 Process Quality

Quality principles cannot only be applied to the manufactured product but also to the manufacturing processes. A manufacturing process inherits a specific order of transformation activities alongside temporal and spatial dimensions with a defined input and output. The quality of manufacturing processes is determined by the compliance with criteria for order, time, place, input, and output (Kreutzberg, 2000; Wuest 2015).

Process quality determines the product quality (see Figure 9.2), given that the entire manufacturing chain and product/process design are capable of meeting the requirements (Brinksmeier, 1991; Jacob and Petrick, 2007). When we execute a manufacturing process repeatedly with the exact same parameters, we will find a certain degree of variation of the input parameters of the individual products even when we deal with state-of-the-art manufacturing. This in turn can influence the process quality and thus the product quality (Taguchi, 1989; Yu and Wang, 2009; Wuest, 2015).

Achieving a high standard of process quality in manufacturing is a core objective of the manufacturing enterprise. In the following, some of the most commonly used methods and tools for quality management, monitoring, and control are introduced.

9.2 Basic Tools of Quality

Quality is a very broad topic and has multiple facets. In manufacturing, we have to consider both product quality and process quality. In the following sections we will introduce selected tools and methods that are widely used by industry. While not a comprehensive list, which would require its own dedicated book, we believe the selection provides a solid understanding of the range and variating emphasis of today's quality programs.

9.2.1 Ishikawa Diagram

The Ishikawa diagram stands out among similar tools due to its ease of use and structured approach that encourages and supports the users to develop solutions that address the underlying problem and not only symptoms of said problem. Common application of the Ishikawa diagram is product and process design problems as well as quality problem prediction. The Ishikawa diagram is also known as Cause-Effect diagram or Fishbone diagram due to its iconic shape (Figure 9.3). There are different variations available, for example, often the older variations with "man" (or "manpower") instead of "employee" is still widely used under the "6Ms" classification.

A fishbone diagram is a great tool for brainstorming certain situations in a proactive manner and to focus the situation on the problem at hand. To build an appropriate fishbone diagram, we often start by thinking high level connections. Let us consider an issue of print

FIGURE 9.3 Ishikawa diagram (fishbone diagram).

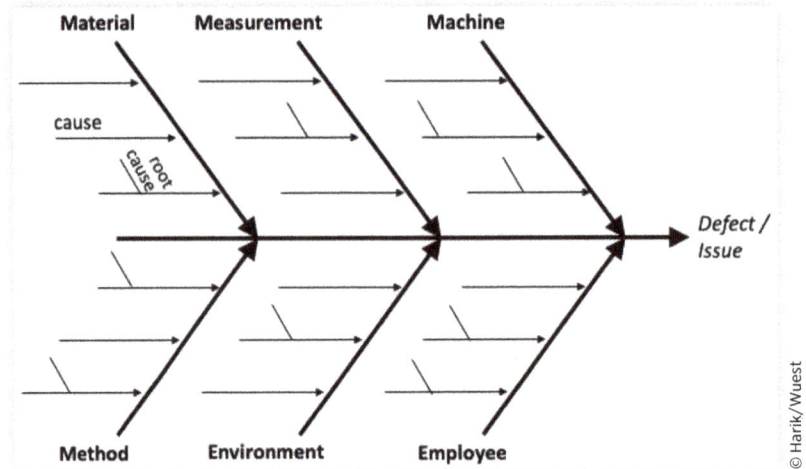

quality issues with a new 3D printer. The first level can include the slicing tool, the printer CNC control, the nozzle, the filament, the part designs, and such. The first level constitutes an aggregation of potential issues. Then, we start developing each level with second level-related problems. Let's take the nozzle and start analyzing what can cause the effect of "bad print quality." Nozzles are chambers in charge of melting the filament and to extrude it in the right quantity at the right time. Temperature, control, feed, extrusion, and other elements can be the second-level problems faced when handling extruder nozzles.

The final diagram would present an overall picture that highlights in a very straightforward manner all the root causes that can have a very specific effect, thus enabling people to explore potential solutions. Often, teams can start looking at each aggregation to tackle issues in a concentrated fashion. Another tool that aims at identifying the root cause of problems is the "5 whys" technique that digs deeper by asking more detailed why-questions in each iteration.

9.2.2 **Pareto Chart**

Pareto chart, sometimes referred to as the 80-20 chart, stems from the concept that most of the time 20% of events are the root cause for 80% of the quality issues (Figure 9.4). Let us consider an example to better illustrate the concept. Imagine the quality defects that can arise in composite manufacturing of Automated Fiber Placement (AFP) parts. The different types of defects that can be expected are approximately 16–17 defects, and they are listed in Chapter 8.

FIGURE 9.4 Pareto chart showing how 20% of events are the root cause of 80% of quality issues.

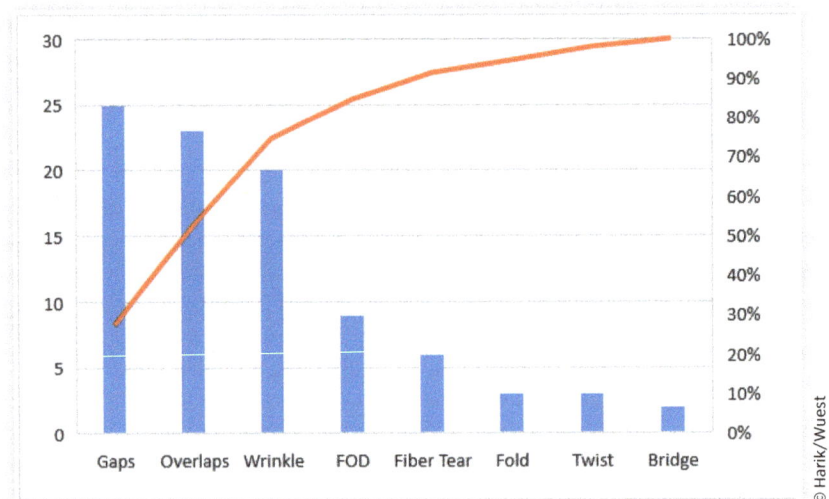

In practice, three of these defects being gaps, overlaps, and puckers/wrinkles are the most common and represent the majority of AFP defects.

The Pareto concept is also commonly used beyond the above-described quality application. As a student, you might have heard that completing a team assignment to 80% takes 20% of the time, and in order to achieve a 100% completion the group has to invest 80% of the time. While this is unfortunately used at times to justify incomplete or unsatisfactory work, in some situations this is a helpful concept to keep in mind when you are overwhelmed with too many different projects to effectively manage your time resources.

9.2.3 Failure Mode and Effects Analysis

The Failure Mode and Effects Analysis, or FMEA in short, is a systematic approach to improve the overall quality of a process, design, or system functions. The basic premise of FMEA is that quality cannot be checked but has to be actively created. An FMEA team is ideally interdisciplinary and approaches quality with the goal of defect prevention. While stemming from the design domain, today, FMEAs are used in a variety of areas including but not limited to process management, software, system engineering, production, logistics, and maintenance.

A FMEA follows a 10-step process (see Figure 9.5): *Step 1* requires us to assemble an interdisciplinary team and define the FMEA object (system and/or sub-systems). In *step 2*, we list all possible failure modes and effects for the focal system without any judgment on probability or risk. In a *third step*, we rank the *severity* following these criteria (1–10):

- Unlikely that the failure will have a significant effect on the product or process (1)
- Marginal failure – failure effects customer only marginally and the system will only be marginally affected if at all (2–3)
- Moderate failure – leads to dissatisfaction of the customer. The system and its functions are impacted and it will lead to repair or warranty claims (4–6)
- Significant failure – leads to enraged customers and the system and/or sub-system is not functional. However, safety and policies/laws are not impacted (7–8)
- Extreme failure – safety and/or policies/laws are impacted (9–10)

The *fourth step* then lists potential causes that may lead to the listed failures. In a directly associated *step 5*, we rank the likelihood or *occurrence* of the causes following these criteria (1–10):

- Unlikely, no problems (1)
- Unlikely – however, check design and/or process to avoid cause (2–3)
- Likely – It is known from similar cases that these causes emerge. Activities to address causes need to be started (4–6)

FIGURE 9.5 FMEA process.

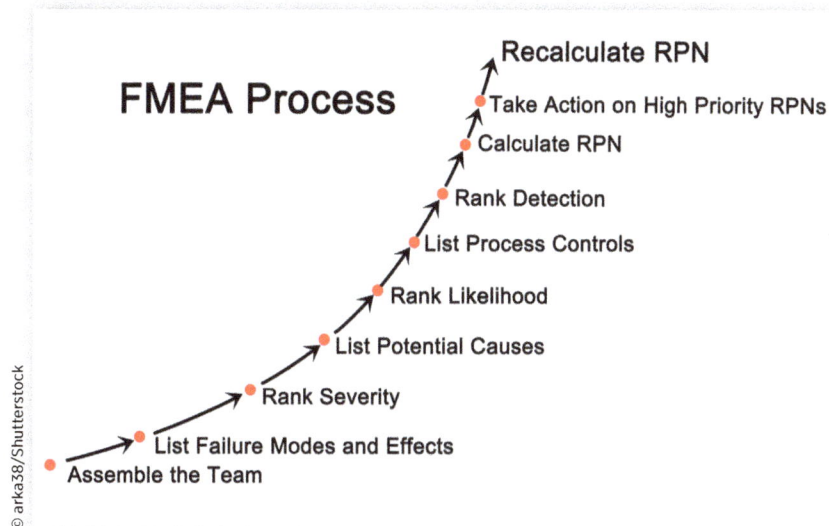

- Regular – design and/or process are known to cause problems and addressed holistically (7–8)

- Extremely likely – the design principle and/or process need to be critically assessed and completely overhauled (9–10)

In the following *step 6*, we collect and list current process and quality controls (*detection*) and rate them in the next step (*step 7*) following these criteria (1–10):

- Safe – failure will directly be caught in the next process step (e.g., functional test) (1)

- High – obvious failure, automatic 100% inspection possible (2–3)

- Moderate – traditional inspection and quality control such as attributed random sampling may catch it (4–6)

- Very low – non-obvious failure, requires visual or manual 100% inspection (7–8)

- Unlikely – failure is not or cannot be identified using inspection (e.g., material properties requiring destructive measurement) (9–10)

We then calculate the Risk Priority Number (RPN) for all possible failure modes in *step 8* by multiplying the three previously derived quantified measures (max. RPN = 10*10*10=1,000) and create a ranking. Then we take action on the highest-ranking failure modes, gradually moving down the list as appropriate (*step 9*). After each activity, we go through the FMEA process again and recalculate the RPN to create a new ranking of the priority failure modes (*step 10*). That way, we continuously reduce the possible causes for quality problems of our system and are constantly aware of potential risks.

Generally, a FMEA is conducted using a table format. There is a huge variety of different designs available; however, the principle setup remains the same (see Figure 9.6).

Let's have a quick look at how an RPN can be determined for a drilling process. We will not look at all possible failure modes but only an exercise to illustrate the FMEA process. We are specifically looking at the *FMEA objective* of "drilling a blind hole." For this objective, a possible *failure mode* is "hole not deep enough (Z-axis)." This failure mode can lead to several *failure effects*, including the to-be-inserted pin sticking out too much. As this can lead to problems during the assembly process and impacts the systems functions – therefore we rate this failure effect with a *severity 6* (moderate failure). When investigating what *potential causes* can lead to an improper depth of the blind hole, again several options occur. We focus on the option of a "worn or broken drill bit" as the cause for the failure. This is a rather common problem; therefore, we judge its *occurrence* with a rating of 6 (likely). In a next step we assess the *current process controls* that are in place. In this case, no process controls are employed; hence, we judge the *detection* probability with a 9 as it is currently unlikely that the cause of the failure will be detected.

FIGURE 9.6   FMEA table highlighting 10 steps.

© Harik/Wuest

Now we *calculate the RPN* for this particular line item by multiplying the three ratings (RPN = severity*occurrence*detection): 6*6*9 = 324. Let's assume this RPN of 324 is ranked high on the list and we decide to take action. In this case, we recommend installing in a sensor-based predictive tool wear analytics system that reliably predicts tool wear and failure before it occurs (*what*), thus allowing us to prevent failures such as inaccurate depth of blind holes. We associate responsibilities (*who*) and due dates (*when*) and, once installed and operational, revisit the potential cause to calculate the new RPN. In this case, the severity and occurrence do not change; however, the detection is now rated with a 1 (instead of a 9) as the cause will be caught before it actually occurs. The new RPN in this case is 36 (reduced from the previous RPN 324) and this will place this potential problem rather low on the priority list.

9.2.4 Control Chart

Control charts are precisely what their name inherently indicates: they monitor variables to ensure they are "in control," which mostly means within an acceptable threshold corridor (see Figure 9.7). Sometimes referred to as Shewhart charts, after their originator, they are a group of graphical representations (time series graph) where we have lower (Lower Control Limit [LCL]) and upper limits (Upper Control Limit [UCL]), all the variable data (input/output) and a clear understanding of which variables/conditions do not fall within the expected behavior of the manufacturing process. The possibility to include time stamps and/or stage gates allows to develop insights regarding the progression of process quality over time.

These charts are abundant in both the business and manufacturing domain. Their comparably simple design and low entry barrier lead to several types of control charts with different objectives being adopted. Variations include combinations such as $\bar{x}$ and R, and $\bar{x}$ and s control charts. Each type of control chart visualizes data in a fashion that is most convenient to a specific situation, industry, or process.

In the following discussion we briefly illustrate the context and the formulation to generate a control chart in a manufacturing context, as well as how to interpret the data presented in the chart. When setting up a control chart, the first step is to define the central line (*average*) as a visual reference that allows us to easily detect shifts or trends of our monitored process (time series). We then calculate the UCL and LCL. They enable us to reduce the monitoring effort by only focusing on issues warranting action, such as when the graph exceeds the calculated limits. UCL/LCL define the process variation limits. There is another band that is sometimes used in control charts that describes the customer variation limits, Upper Specification Limit (USL) and

FIGURE 9.7 Control chart visualization.

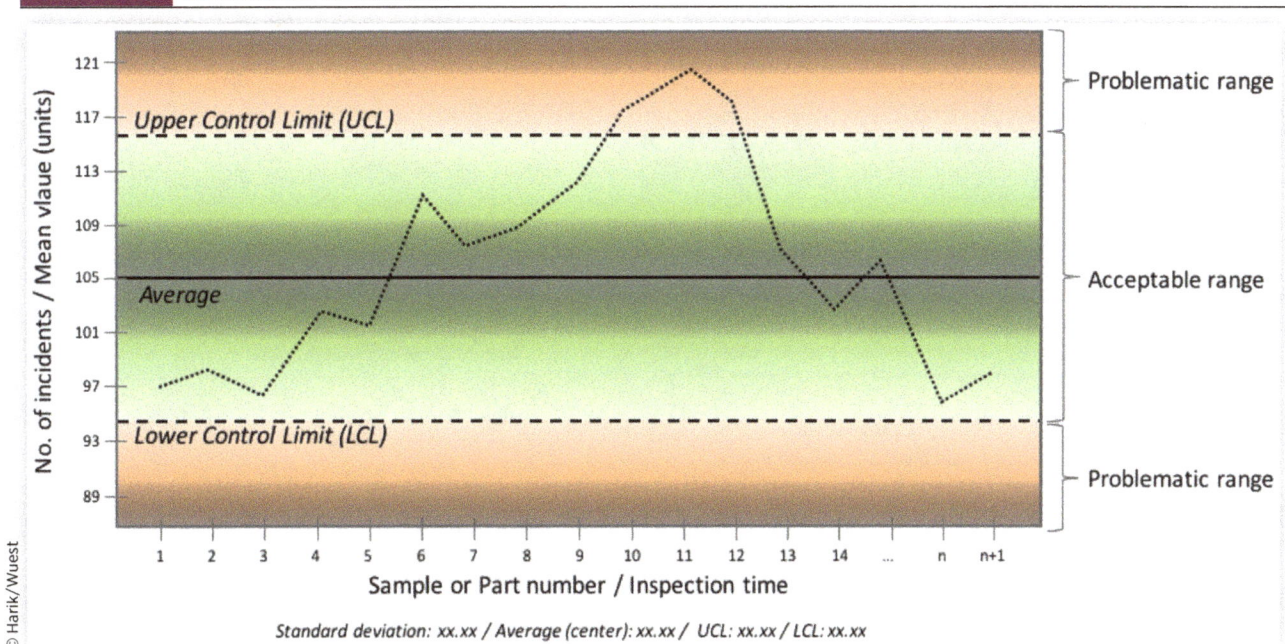

Standard deviation: xx.xx / Average (center): xx.xx / UCL: xx.xx / LCL: xx.xx

Lower Specification Limit (LSL). While UCL/LCL are calculated, there are different approaches how to set the USL and LSL, often dependent on the process that is monitored by the control chart.

A commonly taken approach to calculate the UCL and LCL is based on the estimated standard deviation (EStD) of the process. Once we have the EStD we multiply it by three and add (UCL) and subtract (LCL), the resultant number from the average (AVG). Therefore, we can obtain the following formula:

$$UCL = AVG + 3 * EStD \qquad \text{Eq. (9.1)}$$

$$LCL = AVG - 3 * EStD \qquad \text{Eq. (9.2)}$$

The above calculations for UCL/LCL are rather generic and as such do not reflect the requirements of the large variations we experience in industrial production. There are different variations of control charts for different purposes that mostly reflect how to determine the control limits. A common distinction in the selection process for the most appropriate control limits is the differentiation based on the nature of data represented in the chart: *continuous data* vs. *discrete data*. For continuous data, depending on the sample size we select an appropriate form (e.g., $\bar{X}$). For discrete data we have to distinguish between the possibility of binary and non-binary defects. Binary reflects that a unit is either okay or not okay, whereas non-binary means that there are multiple different defects possible for each unit. For binary systems, we then choose either np-charts (constant sample size) or p-charts (variating sample size). For non-binary systems, we select c-charts (constant sample size) or u-charts (variating sample size).

9.2.5 Quality Function Deployment (QFD)

QFD (also known as House of Quality) is an established method to systematically plan the quality of a product based on customer and/or market requirements. Furthermore, the method provides insights on the required production processes and quality control mechanisms to successfully produce the product in the quality desired. The overall mission of QFD is to always give the customers' voice top priority in all quality-related decisions.

The QFD method follows 11 distinct steps that result in a priority list (ranking) of critical design requirements. Figure 9.8 depicts the principle setup of the so-called house of quality, which is the core of QFD. Furthermore, the 11 steps are highlighted in the quadrant of the house of quality. In the following, we will discuss briefly each step and detail how the house of quality is filled with content to ultimately derive the ranking of critical design requirements (see Figure 9.9). First, we cover the first three steps in a horizontal move, reflecting the *customer-centric dimension*. In *step 1*, we derive the customer requirements of potential customer of our planned product

FIGURE 9.8 Principle house of quality process (QFD).

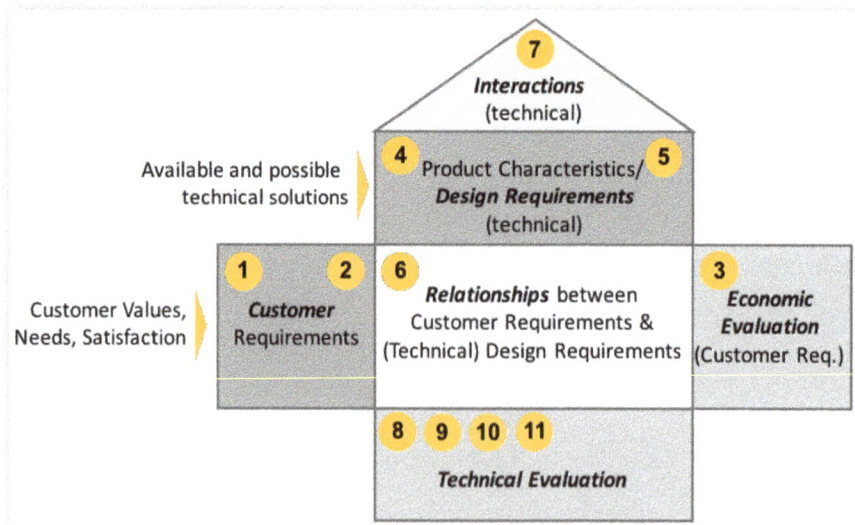

© Harik/Wuest

FIGURE 9.9 Exemplary use of QFD method house of quality.

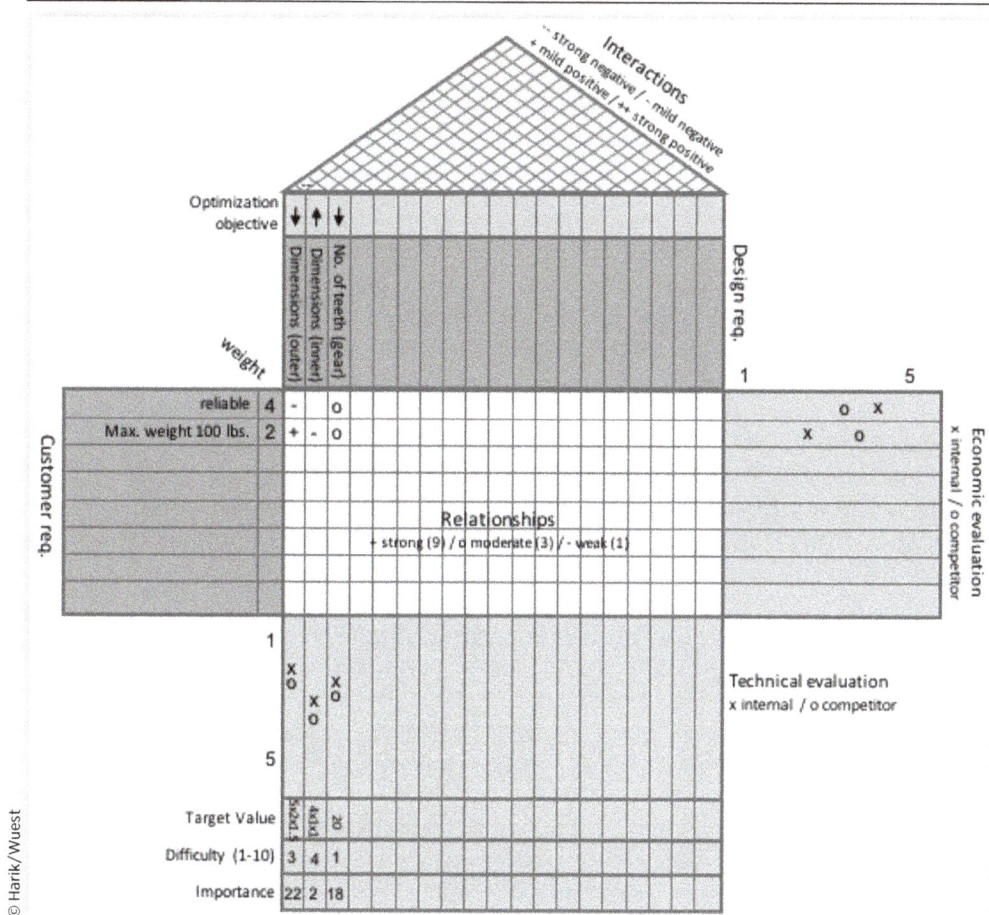

© Harik/Wuest

and associate a weight to each (*step 2*). The *third step* evaluates our own product with the most relevant competitor products on a scale from 1 to 5 from an economic or market perspective.

The *technical or engineering-centric dimension* covers steps 4–11 and can be seen as a downward vertical move. *Step 4* requires us to collect all design requirements that are needed to fulfill the customer requirements or to describe them in a qualitative way. For each of the design requirements, we associate an optimization objective (*step 5*). We try to fit all within three categories: minimizing (downward arrow), maximizing (upward arrow), and required target value (o). In *step 6* we then develop the correlation matrix at the heart of the house of quality where we check the influence of certain design requirements on customer requirements. We associate three different intensities with associated weights: strong (9), moderate (3), and weak (1). In the following *step 7* we investigate interdependencies of the different design requirements for each possible pairing under consideration of the individual optimization objective. In case there is a positive correlation, meaning that the improvement of one design requirement leads to an improvement of the other, we provide either a mild (+) or strong (++) positive rating. On the other hand, when the improvement of one design requirement leads to a decline in the other, the negative rating is indicated by mild (−) or strong (−−). The *tenth step* provides an estimation regarding the technical feasibility of achieving the design requirements on a scale from 1 to 10. In *step 9* we define specific characteristic quantifiable target values for the design requirements. *Step 10* provides a technical evolution of the solution compared to competitor products from an engineering perspective. The final *step 11* derives a quantitative assessment of each design requirement that leads to the final rating that helps to identify the most critical ones to allocate resources and focus appropriately. To get this rating number, each weight (1–10) of a customer requirement is multiplied with the intensity of the relationship in the correlation matrix (9 "strong"/3 "moderate"/1 "weak") and then multiplied for each design requirement (column).

One of the strengths of the QFD approach compared to the conventional approach is the systematic process and the strong emphasis on the market and competitors (Figure 9.9). It can

CHAPTER 9

be applied to whole products or on the component or part level, which makes it very versatile within an engineering environment. In the end, it makes complex design decisions manageable by highlighting the priority design requirements but also, sometimes even more important, identifies the non-critical ones (aka with an empty column).

9.2.6 Total Quality Management (TQM)

TQM is understood to be the pinnacle of quality focus within organizations (Figure 9.10). As the name indicates, it reflects both the importance of quality within the organization and the overall support across all organizational functions. TQM is built on a customer-centric view, a culture of continuous improvement, and includes all employees and processes. Organizations that are truly committed to TQM are rare and often they can locate themselves on a lower quality tier, such as *Quality Management* which focuses on avoiding quality problems across business units; *Quality Assurance* focusing on statistical tools, process control, and quality cost centers; and the lowest tier, *Quality Control* focusing on final inspection, selection, and rework. The higher tiers of course incorporate the tools of the lower tiers in their quality approach as well.

Quality management is often associated with certification based on standards. Those can be industry-specific standards, which are common in established and powerful industrial sectors such as automotive or aerospace. There are too many standards today for us to introduce them all; therefore, we will focus on one standard that is widely used across industries: ISO 9001:2015. This is the latest update of the ISO 9000 Quality Management Family of Standards.

ISO 9001:2015 defines criteria for organizations' quality management systems. It is currently the only ISO 900X standard that can actually be certified. The standard is very versatile and is used by big and small organizations from variating industries (manufacturing, automotive, academia, banking, etc.). Today, there are over one million companies and organizations in over 170 countries certified to ISO 9001.*

While the ISO 9001:2015 standard is universal, there are additions for certain industries that reflect their specific requirements towards quality management. Examples for industries with individual add-on packages are Medical Devices (ISO 13485), Governmental Organizations (ISO 18091), Oil and Gas (ISO 29001), and Software Engineering (ISO 90003).

The ISO 9001 standard defines a number of basic requirements towards Quality Management Systems with regard to process, data management, and transparency/clarity of the organizational

FIGURE 9.10   Total quality management.

* https://www.iso.org/iso-9001-quality-management.html.

and operational structure. Furthermore, it is defined by a strong focus on (internal and external) customer requirements, demands dedication from all levels of the organization (including senior management and C-level), and, most importantly, a continuous improvement process and culture.

Today, there are hundreds of consultants and companies specialized to support organizations during their certification process for every cycle. A certification cycle describes the time after which a new certification is required, and during which improvements have to be documented for the next review. Depending on the performance, the time between review cycles can vary (great performance = longer cycle; weak performance = shorter cycle). Especially transitioning from the previous version, in this case ISO 9001:2008 to the newest version is challenging, and outside help is in high demand.

9.3 **Line Balancing**

Line balancing handles a fundamental task: How to efficiently run a specific job that includes several tasks. To illustrate the intention of line balancing as a productivity measure, we will discuss the most straightforward example.

Imagine a manufacturing process where we drill a hole in a block and, after that, we paint the block using red paint. We have two operators A and B, each assigned to perform one of the functions stated above. Operator A is in charge of drilling the hole, which takes 2 min to perform. Operator B, who is in charge of painting the block, takes 1 min to perform the operation. This scenario describes a worst case in terms of productivity and resource utilization. Operator B, whose task takes 1 min to perform, has to wait for an additional minute to receive the next part with the drilled hole from operator A once done with his task. The result is that operator B is idle 50% of the time, literally wasting time standing around.

Line balancing methods are employed to recreate and redefine the tasks, with the objective to reduce the idle times of operators. Several questions need to be appropriately answered prior to defining the most suitable line balancing method. Do we have synchronous or asynchronous service? Do we have machine reliability issues that we need to account for? What kind of time study was conducted to estimate required times to perform actions? What kind of operator rating is included to assess the line balancing problem?

The above issues each require their own chapter to be properly defined and presented. The objective of this overview section is to caution the reader that we have to address different aspects of the described problem in more detail in order to make appropriate assumptions to resolve the problem at hand.

9.3.1 **Synchronous Service**

Synchronous service processes (*synchronous line*) describe a setup, where individual process steps are synchronized when it comes to workpieces transferring to the next process step. In this system, all process stations (machine tools, assembly units) wait for the slowest process to complete the operation before all workpieces are released at the same time to move to the next station. An advantage of a synchronous line is that there are no buffers required as workpieces are either being processed or in transfer. However, a potential downside of this organization system is that, if not carefully planned, machine tools with faster processing time are idle for, at times, extended periods while waiting for other processes to finish.

9.3.2 **Asynchronous Service**

In contrast to synchronous service processes, asynchronous service processes (*asynchronous line*) release workpieces from the manipulating process as soon as the operation is completed. They are then either transferred directly to the next required process or stored in the warehouse (long term) or used as (in-line) buffers (short term). An asynchronous system generally increases the process utilization (reduction of idle time/load) and allows to address possible bottlenecks (e.g., processes with longer processing times) more effectively. However, asynchronous service processes demand a more careful planning as it involves the strategic use of buffers.

9.3.3 Time Studies

Time studies, sometimes referred to as combined time and motion studies, describe a structured process that aims at observing and measuring the human work process. Time studies originate from the work of the industrial efficiency pioneer Frederick W. Taylor. The better we can anticipate how long a granular sub-process (process-component, e.g., tightening a screw) will take the better we can plan the overall process (e.g., assembling a car door). Time studies can be conducted by directly observing a human operator and measuring work or process increments with a stopwatch, or by other means such as videotaping. It has to be mentioned that time studies are often critically discussed, and in some countries, time studies are only allowed by certified professionals and signed off by labor unions (e.g., REFA certification or equivalent in Germany) due to the perceived potential of misuse. It has to be noted that time studies should be used carefully, and the human operator's work should not be planned and controlled to the millisecond. We can observe a change in perception towards a more "self-organized" workplace design, away from very tightly optimized tact times. While still allowing for optimized planning of the overall process, longer tact times with more variety of tasks in each tact interval are understood to correspond better with human nature and prevent quality-related problems, such as lack of focus, bore out, or burn out.

9.3.4 Operator Ratings

Operator ratings are resembling the act of rating an operator's activity in an attempt to balance the work load across the process. This builds on the understanding that no worker can work highly concentrated on challenging tasks and time pressure for, e.g., an 8-hour workday. Therefore, operator ratings try to quantify activities based on a rating scale (e.g., British Standards Institute Rating Scale). Ratings can range from 0 (no activity) to 100 (full concentration/high frequency). Operator Rating is challenging as it depends on a variety of factors that differ from worker to worker and can only be standardized so much. These factors include (selected): worker being over careful, perceived difficulty of task, unnatural speed of work by operator, etc. In order to conduct the rating, some qualifying conditions need to be met, including worker must be qualified for the task, worker must follow (safety) rules and policies, etc.

9.4 Advanced Inspection Tools and Technologies

In this section, we provide an overview of advanced inspection tools and technologies that are often used to measure and qualify the quality of a certain manufacturing process. While we strive to cover this topic holistically and create a comprehensive list, the sheer number of different tools and technologies in this space renders this task impossible. Therefore, we collected a combination of standout state-of-the-art examples such as light-field camera-based visual inspection and established tools such as roughness measurement systems. The tools and technologies range from simple visual inspection to extensive Computed Tomography (CT) scans to investigate internal deformations of parts. Again, the presented tools are to be understood as a snapshot sample that spans different manufacturing techniques and are not, in any way, an exhaustive list of inspection tools. They offer the reader a first-level understanding of the wide span of inspection tools and technologies.

9.4.1 Traditional and Automated Visual Inspection

Visual inspection, sometimes referred to as Optical Quality Control, is a very common technique to verify part or final product quality. In its simplest form, *traditional visual inspection* is conducted by a human operator that inspects either each part or product as a last step of the manufacturing process (100% inspection), or a statistical sample of the parts or products. This form of visual inspection is limited to quality issues located at the part or product surfaces,

FIGURE 9.11 Automated visual inspection system.

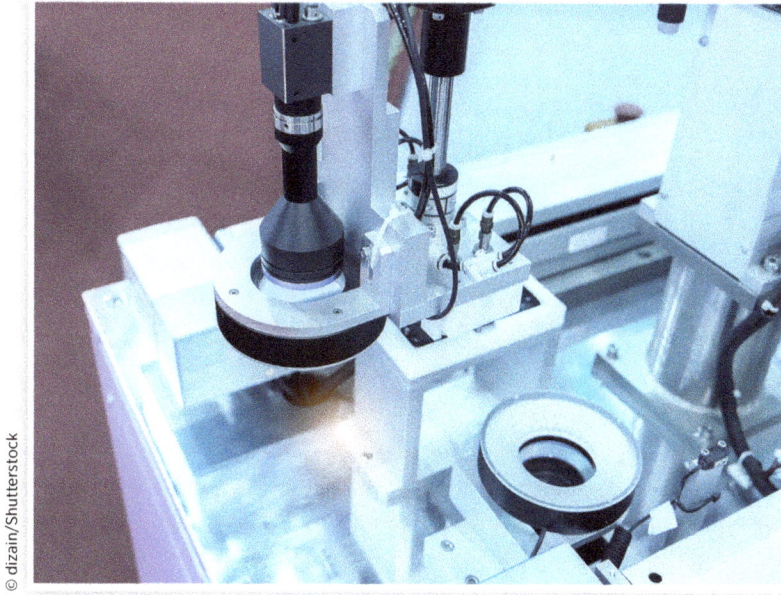

© dizain/Shutterstock

such as scratches or deformations. Internal quality issues, such as residual stress, cannot be identified with this technique. Some companies utilize visual inspection throughout their manufacturing and assembly processes as it is a core skill of each human operator and very flexible and powerful. An example for such a company is premium white goods manufacturer Miele. Miele reverted the automated installation of the main seal of their washing drums back to manual, human operated installation. The main reason for this move was that the human operator, while installing the seal, can inspect the washing drum for scratches and imperfections. For premium manufacturers such as Miele, quality is essential as it justifies their significantly higher price, and as such, investments in quality improvements are seen as a core value of the organization.

While traditional visual inspection is conducted by human operators, today we increasingly rely on camera-based vision systems for *automated visual inspection* (Figure 9.11). Industrial-grade camera systems are capable of taking high-resolution pictures with high frequencies. And to top this off, they can be installed in situ for a 100% inspection at critical sections of the process. Powerful machine learning and AI algorithms allow for real-time analysis of the captured pictures allowing to provide direct feedback to the manufacturing process and quality engineers.

One of the most exciting and advanced technological developments in this space are light-field camera-based inspection systems. Light-field cameras describe camera systems that measure the direction of light in addition to capturing the light through a lens like regular cameras. Therefore, light-field cameras allow for in situ 2D and 3D inspection at high speeds (up to 140 fps)(Weimer et al., 2014). Together with a deep convolutional neural network (DCNN) architecture to analyze and cluster manufacturing defects, these systems present a very powerful system for processes where 100% inspection of quality is required such as pharmaceuticals and medical devices. Weimer et al. (2014, 2016) describe a use case utilizing a light-field camera setup with DCNN to inspect micro parts with less than 1mm in size that are used in the pharmaceutical industry. They successfully showed that the expensive and time-consuming 100% visual ex situ inspection by human, microscopic quality operators can be replaced by a 100% automated in situ visual inspection.

9.4.2 Roughness Measurement

Very often, the intended output of the manufacturing process is a surface specification or a surface tolerance that baselines the departure of the surface texture from a certain nominal geometry. The deviations of the texture from the nominal geometry characterize how much the surface is fit for a specific purpose. Surfaces intended for assembly mating are held to tighter tolerance to ensure fitness, as well as those that are aesthetical (customer facing).

We can compute surface roughness through several formulations. The most common roughness values are

- R_a, which stands for the arithmetical mean deviation roughness, is computed by averaging the peak heights and valleys from the mean line.

- R_{ms}, which stands for the root mean square roughness, is more sensitive than the arithmetic average height to large deviations from the mean line. It is considered as the standard deviation of the distribution of surface heights.

Measurement of roughness can be achieved by simple systems or very complex ones, depending on the needed accuracy. One of the most advanced methods is the Atomic Force Microscopy which has become the de facto tool for imaging surfaces. Using a sharp tip at the end of a cantilever, surface features are measured by the deflection of the tip. This would generate a topographical image.

Other more common measurement systems are Romer arms- and Faro arms-like systems. They act as a 6 degrees of freedom arm that holds either a mechanical tip or an optical one. Mechanical tips are those that require actual touching for the part to capture the measurement. Optical ones use laser scanning like Profilometry to scan the surface from distance.

9.4.3 CT Scan

The prior paragraph introduced external geometrical features, how we assess them and acquire the data. Industrial CT scanning is much like human CT scanning, only implemented to inspect the interior of a part without destruction. CT uses the same technology as the medical field where several waves respond differently to different mediums and as such it produces a 3D image of the solid object's internal features.

The advantages of using CT scanning are numerous: accurate internal dimensions, nondestructive testing, and rapid scanning with high resolutions. CT scanning can be particularly interesting to assess defects in casting and deformative manufacturing. More recently, CT scanning is highly used in composites manufacturing to observe post-cure behavior of composites following the autoclave procedure.

9.4.4 Surface Energy

Surface energy is a characterization of the ability of a surface to adhere/provide energy for adhesion. The best method to characterize surface energy is using the ballistic deposition method that through measurement relevant to a liquid drop and its contact angle can be an indicator of the surface energy.

9.4.5 Double Cantilever Beam (DCB) Tests

DCB tests are well described in the ASTM standard D5528-01. The goal is to create a joining mechanism and then assess, when a load is applied in a specific orientation, how do the two specimens get disassembled. This provides excellent information on the Mode I interlaminar fracture toughness that is an important indicator of expected material behavior.

9.4.6 Scanning Electron Microscopy

Scanning electron microscopy, or SEM, provides detailed surface topographical changes by the use of a focused beam of electrons. SEM is particularly interesting when images of 10k magnification are desired and can be used for interpretation. In the composites manufacturing world, SEM is heavily used to observe fiber exposure after treatment processes and to assess potential adhesion issues.

9.5 **Production Systems**

When thinking about quality control and productivity, we always have to consider the organization of our production system. The production system has a major impact on productivity and also the selection of appropriate manufacturing processes and quality control methods and tools. Figure 9.12 provides an overview of the most common production systems used in industry today and clusters them with regard to product variety and production volume. This section is structured following the three major clusters depicted in the different shades of gray: *Job Shop, Cell or Batch Production*, and *Flow Production/Assembly Lines*. Concluding this section, we have a quick look at selected productivity methods such as Kanban, Push/Pull, and other methods summarized under the umbrella of Lean Manufacturing. Before we delve deeper into the different categories, it is important to understand that the categorization is generalized and the sharp distinction between the bubbles depicted in Figure 9.12 is more of a gray line. In reality, that strongly depends on several factors, such as industry, product specs, materials, complexity, and many more. Furthermore, we have to keep in mind that almost no company acts in isolation and are in some form part of a supply chain or supply network. Therefore, most final products contain different parts that each may have been manufactured in a different production system setup.

In this book, we focus predominantly on discrete manufacturing. Therefore, we continue this focus when covering production systems that are commonly used in industry. We do not cover continuous production-focused systems and their varieties at this point.

9.5.1 **Job Shop**

Job shop production systems are one of the most common setups we find across most industries from small to very large manufacturing companies. While Small- and Medium-sized Enterprises (SMEs) are more likely to organize their operations in a job shop setting, there are large companies that also utilize this setup, e.g., manufacturers of specialty equipment (tunnel drills, machine tools, space crafts, or weapon systems). Job shop production systems are well suited for small-scale production and a large variety of different products. This flexibility reduces the efficiency of production; however, the ramp-up and planning before production can start is reduced as well.

Job shop production systems cover a range of different setups when we take a closer look. Figure 9.13 depicts two common setups, with the workshop layout and the functional layout. The *workshop layout* is generally understood to resemble the production system with the least emphasis on efficiency and layout planning. Some workshop layouts are more or less random and emerged organically by the gradual addition of new machine tools and manufacturing resources.

FIGURE 9.12 Overview of production systems.

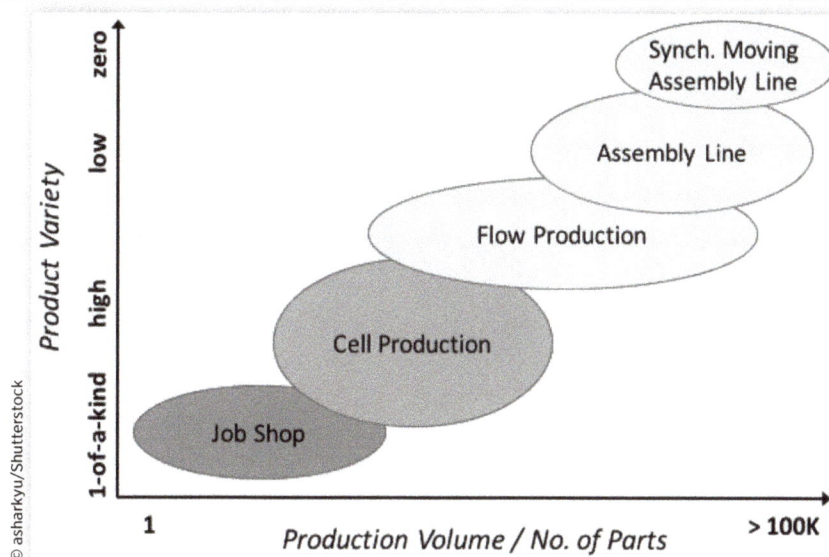

FIGURE 9.13 (a) Workshop and (b) functional layout of job shop production system.

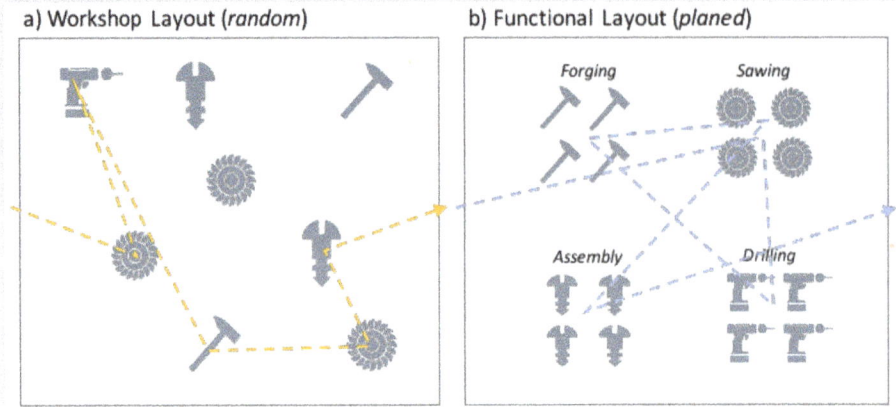

a) Workshop Layout (*random*)

b) Functional Layout (*planed*)

Forging   Sawing

Assembly   Drilling

© Harik/Wuest

Others are limited by the possibilities the facility offers and at least try to arrange the resources following a certain logic (e.g., material flow, noise). We often find a workshop layout in small machine shops or maintenance departments in larger organizations. A more organized and planned variation of the job shop production system is the so-called functional layout. The functional layout arranges manufacturing resources following a certain logic. Often similar resources are grouped together, such as Subtractive Manufacturing tools (such as mills, lathes, and drills), or Hot Working areas (e.g., welding, forging, or casting). That way, the material flow follows the required functions through the facility and the qualified operators have better control over their equipment. From an efficiency standpoint, this setup helps to identify bottlenecks or problematic equipment and thus helps to reduce unnecessary idle time compared to the workshop layout.

Job shop production systems exist in a variety of sizes and setups and can contain very elaborate machine tools or, on the other extreme, are limited to older equipment and hand tools (Figure 9.14). This wide range of manufacturing capabilities leads to a sheer endless variation of products that can be obtained from a job shop production system. A downside of the job shop setup is grounded in its flexibility and rapidly increasing complexity when a certain number of machines are involved. This makes management rather difficult and, furthermore, ownership of the product and its quality is considered problematic. It is difficult to identify bottlenecks and also underutilized machines during production but more importantly while acquiring new contracts/orders. Overall, we can safely assume that a job shop setup is the most flexible and least efficient production system of the three we cover in this book.

FIGURE 9.14 Job shop production system.

© Shutterstock

9.5.2 Cell Production Systems

Cell Production Systems or Cellular Production is a compromise between the flexibility of a job shop layout and the efficiency of a flow production in some sense. In a cellular production system, manufacturing processes and tasks that are frequently used together are grouped in cells. The reasoning for grouping processes together can be product centric, aka all processes needed to manufacture a certain part or module are grouped in a cell. So all processes needed to produce Part A are grouped in Cell A, and all processes needed for Part B are grouped in Cell B. Or we can group processes that are generally used together given the process requirements. This can for example be a turning process (lathe), a drilling process, then a welding process, and a subsequent grinding process.

The cellular setup often also involves the teams or operators conducting the manufacturing operations. This was found to increase ownership and pride in the product and thus ultimately the overall product quality. This is a good reminder that quality is more than just statistical tools and 100% monitoring, but also embedded in the values of the company and its people. An example how cells can be set around teams are premium car manufacturers for their flagship cars, e.g., VW's Phaeton and Bugatti, and Mercedes' Maybach. Both companies developed state-of-the-art facilities for these products where the cell manufacturing philosophy was omnipresent.

A recent example for cell manufacturing is Ferrari's production of V12 engines in their factory in Maranello, Italy. While their V8 engines are manufactured on a synchronized moving assembly line, their premium engine V12 is assembled in a cell setup. Ferrari tasks their best and most experienced engineers to manufacture these high-performance engines using an eight-step cell setup with a takt time of 1 h. Over the term of one full workday, one operator manufactures one fully assembled engine, conducting all required operations. This association of the operator with one specific engine increases ownership, aka pride in the product, and thus the overall quality of these complex systems. Mercedes AMG uses a similar system and the operator puts his or her signature on the engine before delivery. In case of Ferrari's V12 manufacturing, the engine is then directly phased in the synchronized moving assembly line to manufacture the super car. This showcases again that the different production systems are used in combination as needed to achieve the overall objective.

9.5.3 Flow Production and Assembly Line

The origins of *Flow Production* trace back to the principles of mass production, introduced by Henry Ford, and are the defining starting point for the second industrial revolution. Flow production arranges the manufacturing resources in a continuous way to optimize the flow of the parts and products from one manufacturing process to the next. It has to take possible bottlenecks into consideration and, e.g., double, triple, or *n times*, the manufacturing capabilities for a certain process to ensure the product flows continuously and no resources are idle. This requires deep insights in the required times to complete a process and the production logistics. Flow production often utilizes specialty equipment to facilitate the smooth flow of materials and advanced productivity methods, such as pull systems. Within flow production, there are again various different levels, starting with the basic flow production that basically rearranges the assets in a continuous way, all the way to the fully synchronized moving assembly line. We can imagine that planning, ramp-up, and operation of a flow production require significant investment in time, expertise, and money. Therefore, it is generally only employed with high-volume production to justify the initial investment.

The *Synchronized Moving Assembly Line*, most common in the automotive industry, is the most efficient production system we know (Figure 9.15). Literally optimized to reduce idle time and increase throughput, it is a main factor in achieving economies of scale in industrial production. While Figure 9.12 places the Synchronized Moving Assembly Line at the quadrant with low to zero variety, in automotive production, the modular design and Just-in-Time/Just-in-Sequence systems allow the effective use of this production system variant to produce a variety of products on the same assembly line despite millions of different products being produced. However, this can be considered an exception as the automotive industry spent decades of research and optimization to making this work. Automation is

FIGURE 9.15 Synchronized moving assembly line in automotive industry.

© Vereshchagin Dmitry/Shutterstock

utilized significantly in these types of production systems, further increasing the efficiency. Interestingly, in recent years, the takt time of the moving synchronized assembly line in the automotive industry is increasing again. This leads to more and variating manufacturing operations to be conducted by an operator during each step. This change is grounded in the lesson learned that the process quality suffers when operators' tasks are too restricted with little mental or physical variation.

One of the most well-known quality- and efficiency-focused production systems is the Toyota Production Systems (TPS), and its many variations (Mercedes Production System, Porsche Production Systems, etc.). TPS brings together philosophical aspects and applied management practices to optimize production holistically. One key aspect of TPS is that it understands production as a complex socio-technical system, finally giving the human factor the credit it deserves. TPS was not created overnight but was developed over decades in the mid-1900s by Taiichi Ohno and Eiji Toyoda. It is generally considered the basis for today's omnipresent Lean Manufacturing practices. Porsche, who adopted the TPS practices late, was so successful in improving their operations and manufacturing processes that they formed a dedicated consultancy, Porsche Consulting. This branch now helps other companies in various industries to profit from the lessons learned on the shop floors of the automotive industry.

9.5.4 Lean Manufacturing – Productivity Tools for Production Systems

Within production systems, there is a vast array of different tools and methods available that aim to improve the quality and productivity of the system. A rather popular set of methods is referred to as *lean manufacturing* or Lean for short. Lean is popular beyond manufacturing today and stems from the work done by Toyoda and Ohno. The guiding principle behind Lean is to reduce "waste" (*muda* in Japanese) using a systematic approach and different Lean tools (Bicheno and Holweg, 2009). Waste is not only understood as unnecessary activity or use of resources but also referring to an uneven system that should be addressed to be sustainable and productive.

Popular tools associated with Lean manufacturing tools include (selected)

- *Five S (5S)*: Five S stands for a method to organize and operate an effective and efficient workplace. The name originally stems from five Japanese words that exemplify what needs to be done to optimize a workplace: "Seiri" (sort), "Seiton" (straighten), "Seiso"(sweep), "Seiketsu" (standardize), and "Shitsuke" (sustain). Today sometimes the variations *Six S (6S)*, adding the word "safety," or even *Seven S (7S)*, adding the word "spirit," are used.

- *Kanban*: "Pull" production flow in contrast to the traditional "Push" approach. In a Kanban system, an activity is triggered by an actual need not by a pre-scheduled time. A common example for Kanban is the refilling of bolts and nuts at an assembly line.

Once a box of nuts reaches a certain limit (e.g., number of nuts left, eight in a box), called "red zone," an order is triggered to refill the box. Kanban can be done both digitally (aka Automated) and manually using Kanban cards.

- *Control Charts* (see section before).

- *Value Stream Mapping*: The tracing of the flow of value creation in a process aims at assessing the current state of a system; identifying wastes, bottlenecks, and other opportunities for improvement; and defining the desired future state, while taking a number of defining events into consideration. In contrast to Porter's Value Chain, Value Stream Mapping has a clear focus on value adding activities, such as manufacturing within an enterprise.

- *Poka-Yoke*: Poka-Yoke is a systematic approach to design systems, processes, or work areas by anticipating possible mistakes and subsequently preventing them by means of intelligent design. An example for Poka-Yoke in a manufacturing setting is to design a module in a way that it can only be assembled in one, correct way. This can be achieved by using different diameter holes or other features that prevent errors. If done correctly, a Poka-Yoke optimized process reduces the need for later quality inspection and naturally improves quality.

- *Total Productive Maintenance*: Lean tool to build quality into equipment and to improve overall equipment effectiveness.

Lean manufacturing predates the new manufacturing paradigms known as Smart Manufacturing and Industry 4.0 (see Chapter 10). While some them as conflicting initiatives, more and more researchers and practitioners understand them as complementary (Behrendt et al., 2017; Mora et al., 2017). In this age of digital transformation, a new concept emerged that brings the two perspectives together – Digital Lean Manufacturing (Romero et al., 2018). In this view, the new realities of a smart manufacturing environment with a wealth of data, information, and knowledge creates new "digital waste" that need to be controlled in order to create a leaner and more productive operation. Examples for digital wastes can be creation of redundant data.

9.6 Testing Standards

In this section we present the ASTM standards and how to assess which industrial standard is of interest based on specific applications. It is very important to understand that listing all the standards is not possible; however the reader is referred to (www.astm.org) for a complete listing of detailed standards.

An extremely important step in determining the appropriate standard is to understand which level of testing we are focusing on. Certification processes often require testing at different levels:

- *Structure level*: Where the complete and final structure is tested. One can imagine the wing flex test, where the whole aircraft wings are flexed until they break. This test is conducted to ensure safety factors are met without having overdesigned safety factors.

- *Sub-structure level*: Where a portion of the full-scale structure is tested with the goal of investigating a localized problem and/or behavior. Continuing with the example of an aircraft, a sub-structure level test would take the wing box assembly and put it under a certain test, or the famous bird strike test for the engine sub-assembly and others.

- *Component level*: Similar to sub-structure testing, component level testing investigates parts of the overall system. The difference is that components are isolated individual parts such as, to stick with the previous aircraft example, a blade of a jet engine.

- *Coupon level*: Where a representative coupon example of the final structure is actually considered. Dog bone samples are very common with tension/compression and shear test configurations.

Multiple testing procedures can be used complementing each other for the test pyramid such as Open Hole Compression at the coupon level and Uniaxial/Biaxial Loading at the sub-structure level.

References

Behrendt, A., Mueller, N., Odenwaelder, P., and Schmitz, C., *Industry 4.0 Demystified – Lean's Next Level* (New York: McKinsey & Co, 2017).

Bicheno, J. and Holweg, M., *The Lean Toolbox: The Essential Guide to Lean Transformation* (Buckingham: PICSIE Books, 2009).

Brinksmeier, E., *Prozeß- und Werkstückqualität in der Feinbearbeitung*, Fortschritt-Berichte VDI, Reihe 2: Fertigungstechnik (Düsseldorf: VDI-Verlag, 1991), 256.

De Weck, O.L., Ross, A.M., and Rhodes, D.H., "Investigating Relationships and Semantic Sets amongst System Lifecycle Properties (Ilities)," in *Third International Engineering Systems Symposium CESUN 2012*, Delft University of Technology, June 18–20, 2012, 18–20.

Garvin, D.A., "What Does "Product Quality" Really Mean?," *MIT Sloan Management Review* 26, no. 1 (1984).

Jacob, J. and Petrick, K., "Qualitätsmanagement und Normung," in: Schmitt, R. and Pfeifer, T. (Eds.), *Masing Handbuch Qualitätsmanagement* (München: Carl Hanser Verlag, 2007), 101–121.

Kaiser, M.J., "Generalized Zone Separation Functionals for Convex Perfect Forms and Incomplete Data Sets," *International Journal of Machine Tools and Manufacture* 38, no. 4 (1998): 375–404, doi:10.1016/S0890-6955(97)00042-4.

Kamiske, G. and Brauer, J., *Qualitätsmanagements von A bis Z*, 6., Auflage (München: Carl Hanser-Verlag, 2008).

Kreutzberg, J., "Qualitätsmanagement auf dem Prüfstand, Analyse des Qualitäts-managements von Informationssystemen," Dissertation, University of Zurich, Zurich, Switzerland, 2000.

Masing, J., *Handbuch Qualität: Grundlagen und Elemente des Qualitätsma-nagement: Systeme-Perspektiven* (München: Carl Hanser Verlag, 2007).

Mora, E., Gaiardelli, P., Resta, B., and Powell, D., "Exploiting Lean Benefits through Smart Manufacturing: A Comprehensive Perspective," in: Lödding, H., Riedel, R., Thoben, K.-D., von Cieminski, G. et al. (Eds.), *APMS 2017. IAICT*, vol. 513 (2017), 127–134.

Romero, D., Gaiardelli, P., Powell, D., Wuest, T. et al., "Digital Lean Cyber-Physical Production Systems: The Emergence of Digital Lean Manufacturing and the Significance of Digital Waste," *Advances in Production Management Systems (APMS) 2018*, Seoul, South Korea, August 26–30, 2018, IFIP AICT 536, doi:10.1007/978-3-319-99704-9_2.

Sitek, P., "Quality Management to Support Single Companies in Collaborative Enterprise Networks," Dissertation, University of Bremen, Bremen, Germany, 2012.

Taguchi, G., *Introduction to Quality Engineering* (New York: Kraus International Publications, 1989), 263.

Weimer, D., Scholz-Reiter, B., and Shpitalni, M., "Design of Deep Convolutional Neural Network Architectures for Automated Feature Extraction in Industrial Inspection," *CIRP Annals Manufacturing Technology* 65, no. 1 (2016): 417-420.

Weimer, D., Thamer, H., Fellmann, C., Lütjen, M. et al., "Towards 100% In-Situ 2D/3D Quality Inspection of Metallic Micro Components Using Plenoptic Cameras," *Procedia CIRP* 17 (2014): 847-852.

Wuest, T., *Identifying Product and Process State Drivers in Manufacturing Systems Using Supervised Machine Learning*, Springer theses (New York/Heidelberg: Springer Verlag, 2015), doi:10.1007/978-3-319-17611-6.

Yu, T. and Wang, G., "The Process Quality Control of Single-Piece and Small-Batch Products in Advanced Manufacturing Environment," in *16th International Conference on Industrial Engineering and Engineering Management (IE&EM '09)*, Beijing, 2009, 306–310.

10

Smart Manufacturing

In the previous chapters we introduced the manufacturing families, assembly processes, polymer and composite manufacturing as well as Productivity/Quality Control and CAD/CAM. In this chapter, we present Smart Manufacturing which can be understood as a parallel support function that can be combined with all processes and tools previously introduced. Especially Productivity, Quality Control and CAD/CAM are naturally tying in with Smart Manufacturing because they are all heavily relying on manufacturing data.

The manufacturing sector is currently reinventing itself by embracing the opportunities offered by the industrial internet, automation, and machine learning, just to name a few. This development is commonly referred to as the Fourth Industrial Revolution (Industry 4.0) or *Smart Manufacturing*. While Smart Manufacturing can be observed around the globe, there are distinct differences in how it is adopted and embraced among countries, industries, and companies of different sizes. Small- and medium-sized manufacturers are generally understood as the backbone of the manufacturing sector. However, these manufacturers face specific challenges when it comes to the transition toward Smart Manufacturing.

Smart Manufacturing can actually be seen as the cognitive counterpart of automation of physical processes - thus being understood as cognitive automation. While physical automation, e.g., robotic systems, relieve human operators of stressful, dangerous, repetitive, and heavy workloads, cognitive automation attempts the same for mental tasks that are either stressful and repetitive or require significant processing power (Big Data). Overall, Smart Manufacturing circles around data, from data acquisition, analysis, and utilization through visualization.

The chapter is organized as follows: First, we discuss the terminology around Smart Manufacturing, Industry 4.0 and Industrial Internet. In Section 10.2, we present how platforms are today's way of addressing Smart Manufacturing related issues in industrial practice, allowing for modular development and interoperability. Section 10.3 introduces how data analytics and machine learning are at the core of the Smart Manufacturing revolution in manufacturing and presents commonly used algorithms and their applications in manufacturing. Additionally, we provide a selection of educational resources like tutorials and online courses, open access machine learning software tools and methods as well as openly available manufacturing data set repositories. In Section 10.4, we are discussing the utilization of the analytics results in form of visualizations, as well as the increasingly popular concept of the digital twin. In Section 10.5, we include business and operational aspects and extend the Smart Manufacturing principle beyond the shop-floor by introducing Product Service Systems (PSS) and Servitization.

PSS are a data-enabled business model with an increasing presence in manufacturing systems. Section10.6 of this chapter is addressing the issue of cybersecurity that goes hand in hand with deeper penetration of interconnected and data-sharing entities on the shop floor and beyond.

10.1 Smart Manufacturing Terminology

There are many different terms and names floating around describing the principles behind Smart Manufacturing that are often used interchangeably. While there is a certain logic behind this, e.g., geographical or domain origin, this can get quite confusing at times. Before we start to decrypt some of the most commonly used terms related to Smart Manufacturing, here is a (non-comprehensive) list of terms used in this context: Smart Manufacturing, Intelligent Manufacturing, Industrial Internet, IMS, Industrie 4.0, Industry 4.0, Cyber-Physical Systems, Cyber-Physical Production Systems, Manufacturing 2025, Smart Factory, Factory of the Future, Manufacturing Intelligence, Cloud Manufacturing, Cyber Manufacturing, Cognitive Manufacturing, and many more.

In the following sections, we present the core concept and our chosen term Smart Manufacturing in detail and put it in perspective with the theme of this textbook on Advanced Manufacturing.

10.1.1 Smart Manufacturing

Smart Manufacturing describes the convergence of the virtual and real worlds (see Figure 10.1). The three main Smart Manufacturing principles are connectivity, virtualization, and data utilization. *Connectivity* describes the thrust towards integrating all manufacturing resources and enabling the exchange of data and information through the internet. *Virtualization* focuses on the digital representation of information and resources. This can be in the form of a digital twin to replicate the physical system in the virtual environment or as simple as a management dashboard providing a traffic light view of the status of the production line. *Data utilization* highlights the role that advanced data analytics play when it comes to accessing and deriving information and knowledge from the large amounts of available data.

Smart Manufacturing is often used in unison or even as a synonym with Advanced Manufacturing. In our view this is not entirely correct. While the two terms are definitely closely related, used together with a significant overlap in objectives, there are distinct differences. Advanced Manufacturing includes physical manufacturing processes like additive manufacturing (see Chapter 4) and robotics, novel materials, and innovative products. Smart Manufacturing on the other hand focuses on the data side of manufacturing including the acquisition, analysis, and utilization. Some scholars see Smart Manufacturing as a part of the Advanced Manufacturing field (see Figure 10.2a) while others see Advanced and Smart Manufacturing as two sides of the same (manufacturing) medallion (see Figure 10.2b).

FIGURE 10.1 IT/OT integration enabling Smart Manufacturing.

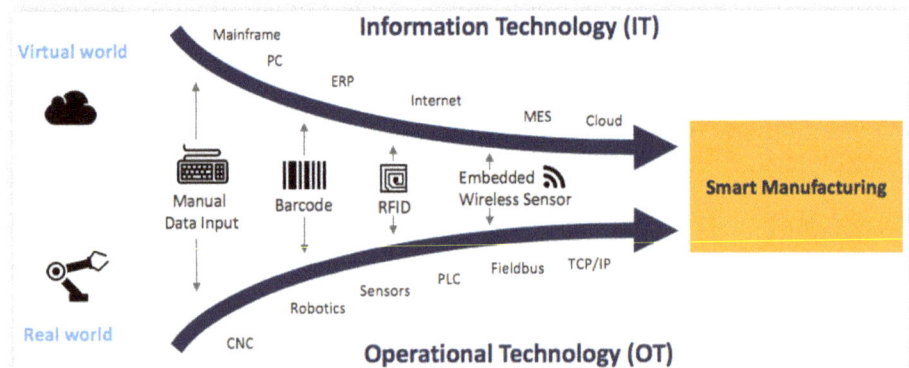

FIGURE 10.2 Distinction of smart and advanced manufacturing.

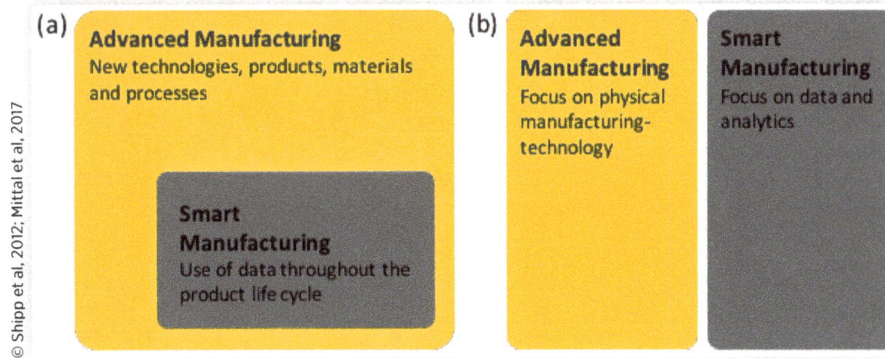

There are several different definitions of Smart Manufacturing available today. Two examples are depicted below:

> *Smart Manufacturing is a data-intensive application of information technology at the shop floor level and above to enable intelligent, efficient, and responsive operations.*
>
> (Wallace and Riddick, 2013).

> *Smart manufacturing marries information, technology, and human ingenuity to bring about a rapid revolution in the development and application of manufacturing intelligence to every aspect of business. It will fundamentally change how products are invented, manufactured, shipped, and sold. It will improve worker safety and protect the environment by making zero-emissions, zero-incident manufacturing possible. It will help keep jobs in this country (USA) by keeping manufacturers competitive in the global marketplace despite the substantially higher cost of doing business in the United States.*
>
> (Chand and Davis, 2010).

While the definitions differ in some capacity and also vary in length and detail, almost all emphasize the core components of Smart Manufacturing as the application of Information Technology/Operational Technology *(IT/OT) at the shop floor level and beyond* to make more effective and efficient manufacturing possible.

The United States recently created two Innovation institutes under the "Manufacturing USA" initiative that address Smart Manufacturing and Advanced Manufacturing topics: the *Clean Energy Smart Manufacturing Innovation Institute* (CESMII) and the *Digital Manufacturing and Design Innovation Institute* (DMDII). Both are unique in the USA in their approach to bring together industry (small to large international manufacturers), service providers, and academic institutions to leverage cross-domain exchange and development. CESMII is focusing on Smart Manufacturing topics with data analytics, OT/IT integration, and service and application platforms at its core. DMDII is focused on Advanced Manufacturing including product design and development as well as the digital integration of associated resources.

The term Smart Manufacturing is used mainly in the USA while the term Industry 4.0 is slowly making its way across the Atlantic from its origins in Germany (Original name in German: Industrie 4.0).

10.1.2 Industry 4.0

The fourth industrial revolution (*Industry 4.0*) is characterized by the introduction of the Internet of Things (IoT) into manufacturing - called the Industrial Internet of Things (IIoT). This transition enables transparency and real-time control over the manufacturing assets, going as far as locating the control of the manufacturing resources in the cloud (aka Cloud Manufacturing).

FIGURE 10.3 Timeline industrial revolutions 1-4.

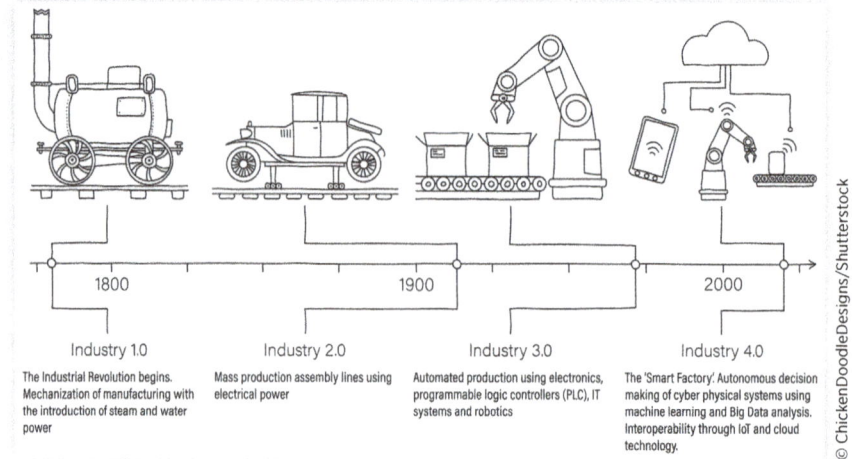

The resulting smart factories are characterized by vertically and horizontally integrated production systems. The vision is that, through highly integrated production industries world-wide, flexible processes that can be adapted quickly enable individualized mass production with the goal to enable manufacturing of individual products (batch size 1) with the same efficiency as large batch sizes. Variants are self-determined through items delivering their own production data to intelligent machines (Shirase and Nakamoto, 2013), which are aware of the environment, exchange information, and control processes in production and logistics by themselves. Data is collected along the entire life cycle in large quantities (Big Data) and stored decentralized to enable local decisions. At the same time, the emerging Industrial Internet platforms allow flexible access and thus the exchange of data with partners and service providers. As data becomes increasingly valuable and manufacturers are becoming aware of data as a resource, there might be new business models in the future where data (or access to data) is sold separately from the products/parts manufactured. In order to realize this vision, elements such as machines, storage systems, and utilities must be able to share information, as well as act and control each other autonomously. Such systems are often referred to as cyber-physical systems (CPS) (Thoben et al., 2017).

The term Industry 4.0 itself emerged from the high-tech strategy from the German government and was first used during the Hannover Fair in 2011 (BMBF, 2018). It builds upon the notion of the previous three industrial revolutions that shaped how we experience manufacturing and actually our world today (see Figure 10.3). Industry 4.0 combines the digitization of manufacturing (e.g., Data Analytics/Big Data, Cloud Computing, and Augmented Reality) with novel manufacturing processes such as additive manufacturing and robotics resembling CPS (see Figure 10.4). In that sense Industry 4.0 resembles a combination of Advanced Manufacturing and Smart Manufacturing.

10.1.3 Smart Manufacturing Technologies

There are many technologies commonly associated with Smart Manufacturing, ranging from machine learning and augmented reality (see Figure 10.5) to holograms. Mittal et al. (2017) provide a comprehensive overview of 38 technologies that are associated with Smart Manufacturing in literature. Many of these reported technologies are similar or closely related, and the authors narrowed the list down by clustering to a set of 11 technology clusters: intelligent control, energy saving/efficiency, cybersecurity, CPS/CPPS, visual technology, IoT/IoS, cloud computing/cloud manufacturing, three-dimensional (3D) printing/additive manufacturing, smart product/part/materials, data analytics, and IT-based production management (Mittal et al., 2017).

It has to be noted that these technologies are all based on published literature and there was no selection or exclusion made beyond this to ensure objectivity. Therefore, technologies like additive manufacturing that we would associate more with advanced manufacturing are included in this list due to the fact that other author reported this as a smart manufacturing technology.

FIGURE 10.4   Industry 4.0 environment connecting people, cloud, manufacturing machines, etc.

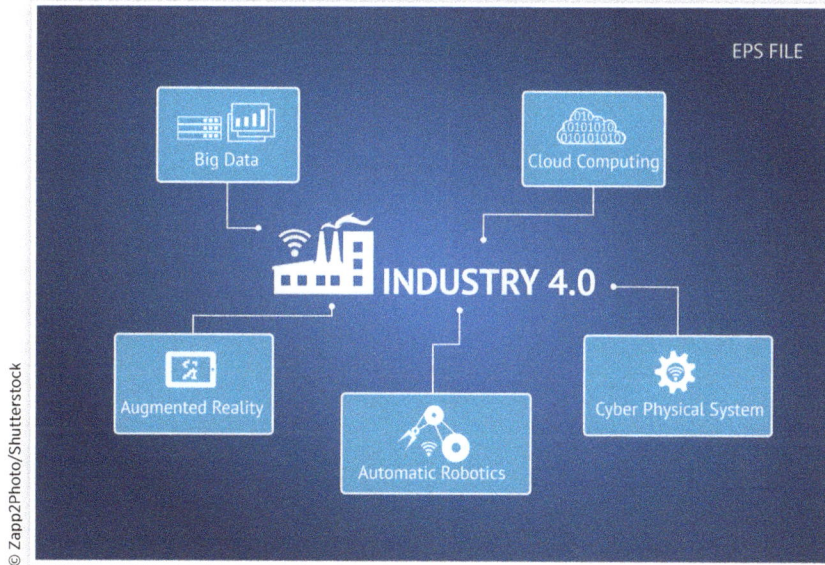

© Zapp2Photo/Shutterstock

FIGURE 10.5   Industrial augmented reality solution.

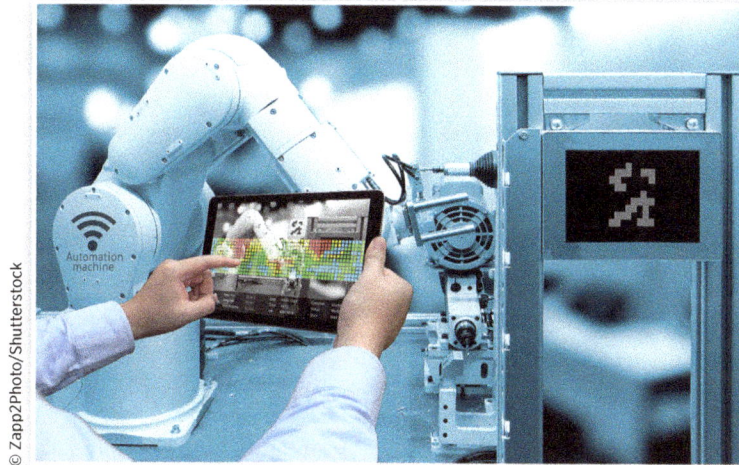

© Zapp2Photo/Shutterstock

10.2 Industrial Internet Platform

Connectivity is a key component of smart manufacturing systems as at its core connectivity in an industrial environment means Industrial Internet (II) or Industrial Internet of Things (IIoT). There is an emerging number of II Platforms available that enable manufacturers to create customized solutions based on their individual requirements without the need to develop a comprehensive software suite or making compromises based on available of the shelf-packages like it was in the past. Developing a new software suite is not only expensive but also limits the flexibility and scalability significantly. Furthermore, manufacturing companies have to invest in either external support for updates and troubleshooting (expensive) or build this expertise in-house which is generally not in line with their core competencies – manufacturing products.

II Platforms offer a solution to this problem by providing a flexible, customizable, and scalable solution which is maintained by the platform provider. Customized modular solutions can be developed and integrated by the manufacturer themselves or third-party service providers for most platforms. While some manufacturers are still concerned with data security when it comes to platforms, there are hybrid solutions available where part of the platform (handling sensitive data – like production data) is managed locally (intranet) and another part is in the cloud allowing for

FIGURE 10.6 IIoT components.

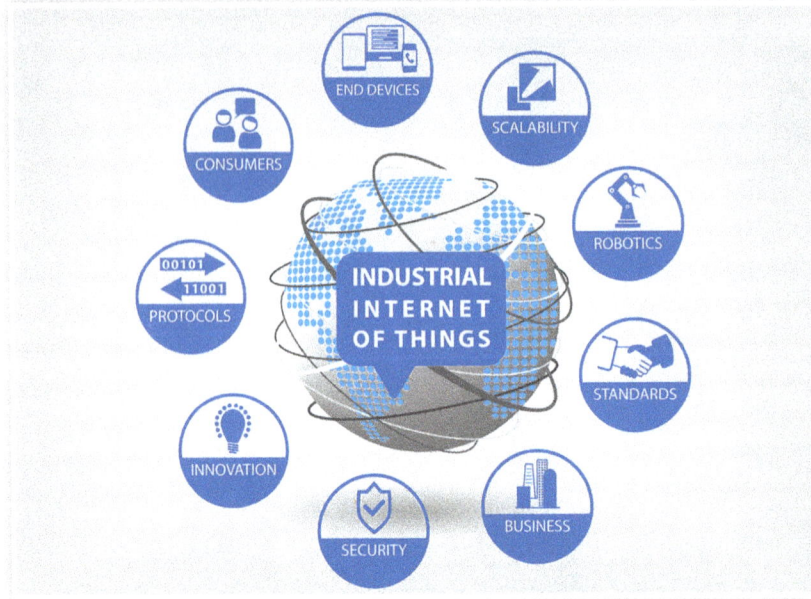

easy integration with the supply chain or other collaborators. Most II Platforms promise to connect all stakeholders (including customers, users, suppliers, service providers) and provide a scalable infrastructure that "takes care of" cybersecurity, data hosting, interoperability, and standards. Furthermore, II Platforms provide a means to connect physical assets (e.g., manufacturing machines and robots) to the virtual environment (e.g., quality analytics, control) (see Figure 10.6).

There is a variety of platforms available today, with PTC's ThingWorx, Microsoft Azure, GE's Predix, Amazon's AWS, and IBM Watson IoT being the most prominent ones. Furthermore, there are some innovative newcomers in the market who provide either specialized platforms for certain companies or industries (e.g., John Deere's Forest Insight) or open-source solutions such as Kaa IoT.

The available platforms all vary in certain aspects. Important differentiators that should be taken into consideration by manufacturers trying to choose a suitable platform for their operations are (selection)

- Business/pricing model
- Openness (e.g., data access, integration of third-party solutions)
- Interoperability/support of (industry) standards
- Integration with/interface to currently deployed enterprise software

10.3 Data Analytics and Machine Learning in Manufacturing

In this section the application of machine learning algorithms and techniques in manufacturing is elaborated. Machine learning has been successfully utilized in various process optimization, monitoring, and control applications in different manufacturing industries. Machine learning techniques were found to provide promising potential for improved quality control optimization in manufacturing systems, especially in "complex manufacturing environments where detection of the causes of problems is difficult" (Harding et al., 2006).

There are many different machine learning methods, tools, and techniques available, each with distinct advantages and disadvantages. The domain of machine learning has grown to an independent research domain over the past decades. Therefore, within this section the goal is to introduce machine learning techniques briefly and provide an overview of how machine learning can be applied to a manufacturing problem. In order to achieve that goal, first, a brief general introduction to machine learning with regard of manufacturing application is presented.

The topic of machine learning firstly gained attention after Samuel (1959) published his paper "Some Studies in Machine Learning Using the Game of Checkers." Since then, not only did the research field of machine learning grow continuously but also it grew more divers. Today machine learning is omnipresent in our daily lives, e.g. through the use of various Google products or certain public transportation systems (Smola and Vishwanathan, 2008).

The research domain of machine learning looks into the practice of preparing computers or artificial systems to act or react to certain events without being specifically programmed to do so (Nilsson, 2005; Smola and Vishwanathan, 2008). Machine learning aims to solve (manufacturing) problems by applying knowledge that was acquired from analysis of (data of) earlier problems of similar nature (historical data) to the to-be-solved problem (Priore et al., 2006). This capability is desirable or may be even necessary in some cases for various reasons.

We experience an increasing availability of large amounts of data which is often referred to as Big Data (Lee et al., 2013). The availability of, e.g., quality-related data in manufacturing offers potential to significantly improved process and product quality. However, it has been recognized that much information can also propose a challenge and may have a negative impact as it can, e.g., distract from the main issues/causalities or lead to delayed or wrong conclusions about appropriate actions (Lang, 2007). Overall, it can be safely concluded that the manufacturing industry has to accept that in order to benefit from the increased data availability, e.g., for quality improvement initiatives, manufacturing cost estimation and/or process optimization, and better understanding of the customer's requirements, support is needed to handle the high-dimensionality, complexity, and dynamics involved (Davis et al., 2015; Wuest, 2015; Loyer, et al., 2016).

New developments in certain domains like mathematics and computer science (e.g., statistical learning) and availability of easy-to-use, often freely available (software) tools offers great potential to transform the manufacturing domain and their grasp on the increased manufacturing data repositories sustainably. One of the most exciting developments is in the area of machine learning (including data mining, artificial intelligence, knowledge discovery from databases, etc.). However, the field of machine learning is very diverse and many different algorithms, theories, and methods are available. For many manufacturing practitioners, this represents a barrier regarding the adoption of these powerful tools, and thus may hinder the utilization of the vast amounts of data increasingly being available.

In the following we present an overview of the machine learning in manufacturing domain, machine learning algorithms and the process of how they are applied to manufacturing analytics problems. Then we will briefly discuss application problems and domains within manufacturing where machine learning was successfully applied before we provide you with additional resources to explore this fascinating and important field further if you are interested in a deep dive.

10.3.1 Machine Learning Algorithms and Techniques

Machine learning problems in manufacturing can be associated with two prediction goals that influence what algorithm and/or technique is most suitable to create the prediction model:

- *Classification*: Classification describes the goal of training a machine learning model that is capable of predicting the correct discrete class label for a product/part (or process outcome). A very common, non-manufacturing related classification problem is the identification of Spam emails. Classification can be further distinguished in "binary" or "multi-class" classification problems. Binary problems describe the goal of associating a product/part to one of two classes. This is very common in manufacturing, e.g., for quality predictions "ok" vs. "not ok." Multi-class classification describes classifiers that associate products/parts to more than two classes. An important classification problem for classifiers is pattern recognition which is used, among other areas, for image recognition. Image recognition is becoming increasingly important for manufacturing applications, e.g., to identify products and/or for contactless, in situ (surface) quality inspections.

- *Regression*: Regression describes the goal of training a machine learning model that is capable of predicting future quantitative values for a process/product (e.g., yield, weight,

FIGURE 10.7   Structuring of machine learning techniques and algorithms (Wuest et al., 2016).

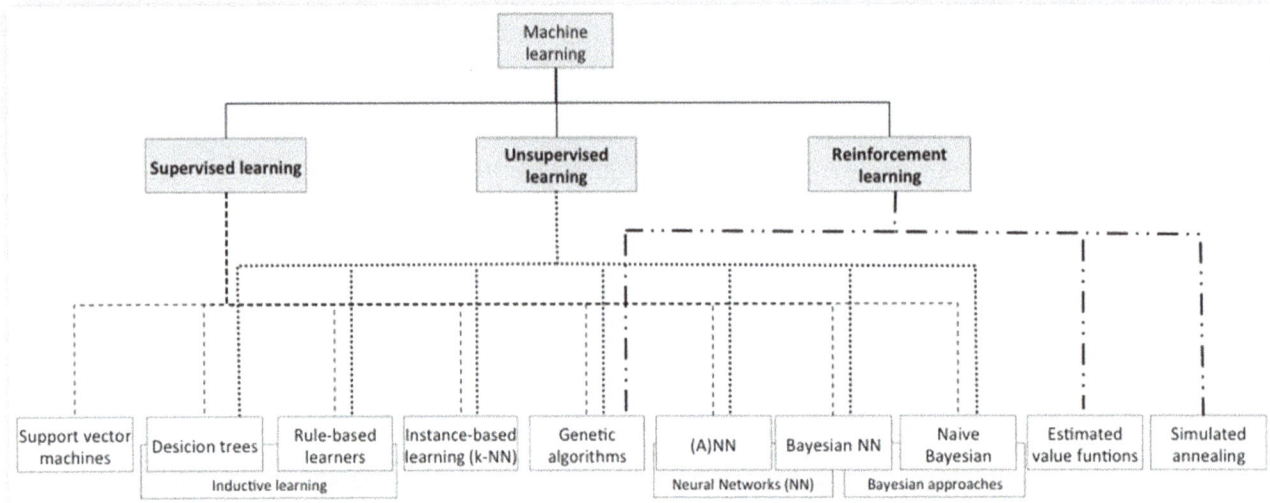

cost, etc.). There are several variations of regression problems that occur regularly in manufacturing. For example, in manufacturing, the time stamp of the input data is often relevant to the output prediction, which makes these problems time-series forecasting problems, a subset of regression models.

There are three main classes of Machine Learning algorithms (see Figure 10.7). Monostori (2003) describes these three major machine learning classes as follows:

- *Reinforcement learning*: less feedback is given, since not the proper action but only an evaluation of the chosen action is given by the teacher.

- *Unsupervised learning*: no evaluation [label] of the action is provided, since there is no teacher.

- *Supervised learning*: the correct response [label] is provided by a teacher.

In a manufacturing environment we use predominantly supervised machine learning. A major reason for this is the availability of domain experts (designers, manufacturing engineers, quality engineers) and, consequently, the possibility of using labeled data for model training. Labeled data means, for example, the classification of parts by the quality control engineering in "ok" or "not ok." This label together with the process and product data (e.g., process parameters, quality measures, and sensor data) accumulated for each individual part builds the basis for the data-driven classification or regression machine learning mode.

10.3.2 Machine Learning Process

In this section we take a high-level view on the typical application process from analytics problem identification - aka What do we want to understand better? – to the fully trained machine learning model that allows us to address the analytics problem in full or in part.

Based on a given problem, the required manufacturing *data is identified, acquired (measured)*, and pre-processed. It is important to remember that we can only analyze what we measure. This means that only data points that are actually measured and included in the data set will be reflected in the resultant model and prediction. For example, in case we want to predict the wear of our cutting tool in a turning process but do not measure the feed rate and store the feed rate data, it is very likely that our prediction model is not very accurate. Due to missing a critical input, the trained model is not equipped to reflect reality accurately.

Pre-processing is an essential step and deserves full attention. Machine learning experts agree that *pre-processing* is key to achieve a good analytics outcome; some even go as far as putting more weight on pre-processing of the data than on choosing the most suitable algorithm regarding impact on the prediction outcome. Pre-processing can include a variety of things such as normalizing and filtering the data, outlier detection and elimination, replacing missing values, addressing unbalanced representation in the training data, etc. Especially the latter is common in a manufacturing environment. Naturally, for an operation to stay in business, the number of "good" (ok quality) products far outweighs the number of non-ok, or scrap, parts. Which is good for the manufacturer's business is not optimal of the analytics task at hand as this leads to a bias in the trained model. There are different methods to addressing.

An important aspect is the ***definition of the training set*** as it influences the later classification results to a large extend. Even so in Figure 10.8, it appears that the ***algorithm selection*** is always following the definition of the training data set, the definition of the training data also has to take the requirements of the algorithm selection into account. This is to some extend also true for the identification and pre-processing of the data as different algorithms have certain strength and weaknesses concerning the handling of different data sets (e.g., format, dimensions, etc.).

After an algorithm is selected, the ***machine learning model is trained*** using the training data set. In order to judge the ability to perform the targeted task, e.g., to identify poor quality products early in the process, the trained algorithm is then evaluated using the evaluations data set. Depending on the performance of the trained algorithm with the evaluation data set, the parameters can be adjusted to optimize the performance in the case the performance is already good. In case the performance is not satisfactory yet, the process has to be started over at an earlier stage, depending on the actual performance. Reliant on the application case, a rule of thumb is that 70% of the data set is used as a training data set, 20% as an evaluation data set (in order to adjust the parameters – e.g., bias), and final 10% as a test data set; however in practice often a 70% (training data) and 30% (test data) split is utilized (Wuest, 2015).

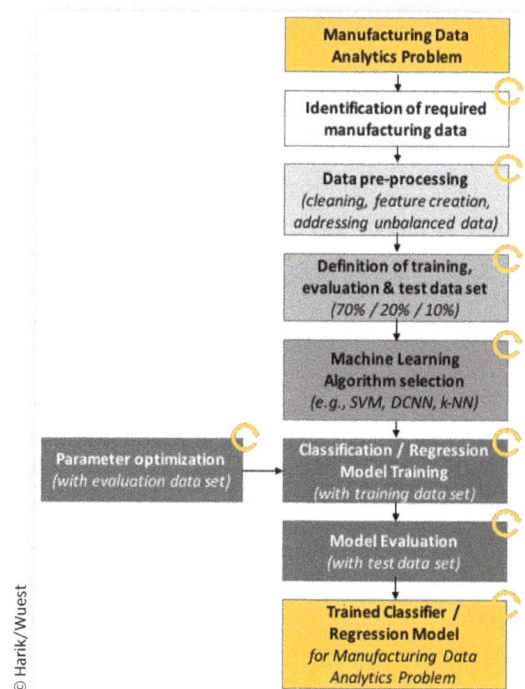

FIGURE 10.8 Generic process of applying supervised machine learning in manufacturing.

10.3.3 Machine Learning Applications in Manufacturing

We learned in the previous section that mainly supervised learning is used in manufacturing. Therefore, most of the applications presented in the following reflect supervised machine learning algorithms like Support Vector Machines. However, there are also increasing applications in manufacturing for semi-supervised machine learning/reinforcement learning models and also unsupervised models like Deep Machine Learning, especially in the field of quality inspection.

Today, machine learning is applied across different manufacturing domains and industries. Common application areas are

i. Quality monitoring and control
ii. Scheduling
iii. Process control (e.g., reinforcement learning for control of laser welding)
iv. Predictive maintenance/condition monitoring
v. Supply chain forecasting; (vi) Optimize manufacturing operations
vi. Asset management
vii. Inventory optimization
viii. Energy optimization
ix. Procurement
x. Testing and calibration of equipment
xi. Image recognition

xii.   Time series forecasting

xiii. Fault diagnosis

xiv.   Tool wear prediction

In order to put some numbers behind the many claims around machine learning applications, McKinsey found that machine learning will lead to 10% reduction of annual maintenance costs, up to 20% downtime reduction, and 25% reduction in inspection-associated costs (McKinsey, 2017).

10.3.4 Machine Learning Tools and Data Sets

In the field of Data Analytics and Machine Learning, there are many resources freely available. A more in-depth overview of Machine Learning in manufacturing is available open access in Wuest et al. (2016). These resources include educational materials like tutorials and instructional videos, online lectures (MOOCs), but also machine learning software tools and a variety of data sets to develop your skills. In the following, we introduce a few popular resources.

Tutorials and online lectures:

- Coursera Machine Learning (https://www.coursera.org/learn/machine-learning)

- Coursera Neural Networks for Machine Learning (https://www.coursera.org/learn/neural-networks)

- edX Machine Learning (https://www.edx.org/course/machine-learning-columbiax-csmm-102x-3)

Software tools:

- WEKA (https://www.cs.waikato.ac.nz/ml/weka/) (free)

- R (https://www.r-project.org) (free)

- Python (https://www.python.org) (free)

- Knime (https://www.knime.com) (free)

- Rapidminer (https://rapidminer.com) (educational version available)

- MatLab (https://www.mathworks.com) (often covered by institutional subscription)

Data set repositories:

- University of California Irving Machine Learning Repository (https://archive.ics.uci.edu/ml/index.php)

- NASA Prognostics Data Repository (https://ti.arc.nasa.gov/tech/dash/groups/pcoe/prognostic-data-repository/)

- Kaggle (https://www.kaggle.com/datasets)

10.4 Manufacturing Visualization and Digital Twin

In a Smart Manufacturing-ready factory, we now have (remote) access to our manufacturing process data in real time, and our analytical models help us navigate the dynamic environment to help us increase efficiency and effectiveness of our production. However, at this point, the output of the analytical models might be pretty much either cryptic numbers or statistical curves that do not directly relate to the task at hand for the operator on the assembly line. In order to make these important insights operational for the different stakeholders of the manufacturing process chain, we need to make sure they are communicated in a way that is supportive and not adds an extra burden on the operator. That is where visualization comes into play.

Because of the way the human brain processes information, using charts or graphs to visualize large amounts of complex data is easier than pouring over large spreadsheets or

FIGURE 10.9 Digital twin of car and manufacturing resources.

© Alexander Tolstykh/Shutterstock

documents of raw data. Generally, data visualization has two main components, data content and the graphical representation of the data. Data visualization is classically used to support different stakeholders to better understand data and underlying information through viewing it from different perspectives using appropriate visuals (Mizuno et al., 1997). Data visualization, used right, can be a quick and easy way to convey complex concepts in an effective manner. In essence, data visualization is more storytelling than simply displaying data. In a manufacturing context, data visualization is understood to be a major research issue within a smart manufacturing environment (Thoben et al., 2017). Data visualization can be used by manufacturers to steer their manufacturing processes effectively and efficiently, e.g., by convening complex simulation results based on slight process parameter adjustments to the process owner as decision support.

In manufacturing, we often face the challenge to design, manufacture, and support complex products that are in operation far from the experts, e.g., a jet engine on a plane over the Pacific while the designers and engineers sit in the Rolls-Royce factory in the UK. In such cases the so-called Digital Twin offers a solution that corresponds with the 3D thinking of most engineers and designers (Figure 10.9).

The digital twin is a virtual representation of a physical product that allows the engineer to analyze, manipulate, and simulate the physical equipment in the virtual space from afar. The virtual model is reflecting the geometry and dimensions of the physical product and augments those with real-time sensor data that is fed from the physical product in operation (Figure 10.10).

FIGURE 10.10 Digital twin of jet engine.

© Gorodenkoff/Shutterstock

The definition of the digital twin describes "the vision of the digital twin itself refers to a comprehensive physical and functional description of a component, product or system, which includes more or less all information which could be useful in all – the current and subsequent – lifecycle phases" (Boschert and Rosen, 2016).

Another virtual representation of a physical product is the so-called Product Avatar. The product avatar can be understood as a digital counterpart or set of digital counterparts which represents the attributes and/or services of a physical product towards the different stakeholders involved in its life cycle. This means a product avatar presents different interfaces and delivery channels depending on who uses it and how. Stakeholders such as owners, designers, and service personnel may interact with the product avatar, e.g., via dedicated desktop applications, web pages, or mobile "apps" tailored to their specific information, service, and interaction needs (Wuest et al., 2015). In contrast to the Digital Twin, the product avatar is more abstract and focused on stakeholder-specific information delivery while the Digital Twin is mostly a more complete representation of the physical system.

There are several examples for the use of Digital Twin. Most of these examples resemble complex equipment that requires a certain expertise to operate and/or be maintained. The prime example is the GE jet engine Digital Twin, but other example includes Wind Turbines, agricultural machinery, and leisure boats. On the shop floor, increasingly individual, expensive high-precision machine tools include a Digital Twin in the virtual space but also some prototypical examples of Digital Twin for whole factories including several complex machine tools, robots, and storage are being developed.

10.5 PSS and Servitization

In this section we will look at an increasing trend that is impacting how products are designed and manufactured – Servitization and PSS. These concepts are extending the traditional focus of manufacturers beyond the point of sale and connecting the products' use phase back to the designer, and offer additional revenue potential throughout the life cycle of the product and satisfy the customers demand for easy upkeep and being environmentally friendly. This is interconnected with many of the technologies and concepts we discussed in this chapter, like IoT, to ensure access to the product during the use phase and PLM to organize the information and data along the life cycle as well as the digital twin to manage the equipment remotely. Therefore, we consider this as an increasingly important part of smart manufacturing and Industry 4.0.

10.5.1 Servitization

An early definition of servitization is provided by Vandermerwe and Rada and states that *servitization is* "the increased offering of fuller market packages or 'bundles' of customer focused combinations of goods, services, support, self-service and knowledge in order to add value to core product offerings" (Vandermerwe and Rada, 1988).

In this definition, services are bundled as add-ons with physical products to deliver additional value to the customers and users. The main benefits of servitization are differentiation against competitors, hindering competitors to offer similar product-service bundles and increasing customer loyalty. Servitization is mainly customer driven, and the availability of mobile data connectivity and new technologies, especially IoT and handheld communication devices (like smartphones and tablets), are fueling this development. In the beginning of the century, over 60% of companies trade at the NY stock exchange offering additional services with their products. Examples are automotive Original Equipment Manufacturers (OEMs) who offer financial services to help customers finance their cars. Of course, they try to make additional profit by providing these services. In some cases, this is successful, e.g., the mentioned financial services, in others, the services are a necessary addition to the physical product and do not offer a profit by themselves. This might however change over time. An example for this is Apple with its iTunes which was at the beginning mainly a music management system offered as a service to drive iPod sales. Over time, the revenue from music sales on iTunes surpassed the revenue made from sales of physical iPod products.

Overall, servitization is a concept that emphasizes the increasing focus of manufacturers on the use phase of their products beyond the point of sale, and how additional services can be offered by themselves or third parties to make additional revenue or improve the customer experience through added value.

10.5.2 Product Service Systems

PSS are following the same line of argumentation with servitization in extending the focus of manufacturers beyond the point of sale. However, PSS are more than mere bundles of services with physical products but emphasize the integration of both with a strong PLM focus and also feature adapted business models to facilitate the changing realities.

The best way to understand PSS is through actual use cases. In the following, we will illustrate two different use cases, one you might have been exposed to yourself recently and one more complex and set in an industry we continuously refer to in this book – aircraft manufacturing:

A commonly used example for a PSS is industrial copiers. Earlier, copiers were purchased by companies, universities, and offices around the world. They also purchase toner and paper and took care of the maintenance. Recently, there is a shift in the business model behind these machines we use every day. Many customers do not own the machine that is physically in their office but are under a service contract that guarantees them to being able to print XX copies a day with an uptime of 99.XX%. In some cases, the paper and toner are still managed by the user; in other, more complete service contracts, the maintenance, toner, and paper is all supplied and managed by the PSS operator. The user does not have to worry about anything regarding the machine and the cost is very clear – X cents per printed copy.

Similarly, jet engines today are not purchased by the fleets any longer but leased. A jet engine is an extremely complex product that is continually monitored (in some cases using a Digital Twin) and carefully maintained by expert operators. Therefore, the jet engine manufacturers out of necessity adopted a PSS business model, which in some cases lead to not selling or leasing the jet engine any longer but selling "thrust" (power by the hour). The jet engine manufacturers are focusing on what they are good at, building and minting the jet engine, and the fleet operators (airlines) have the guarantee that their planes are operational at all times. A very important side effect for the jet engine manufacturers following this model is the access to usage data (flight/operation data). This allows them to optimize the design of their jet engine based on real data points from the operation for each customer. In a world where fuel efficiency and reliability are everything, this is a significant competitive advantage.

What we established through these two use cases is that many complex machines and equipment are no longer purchased by the users. The users do not care about owning a milling machine, they care about the ability to produce precise parts within the tolerances required without interruption and downtime. The idea is that the best option to achieve is to have the machine maintained and upgraded by the manufacturer who actually designed and manufactured the machine tool.

PSS are heavily based on access and analysis of data and information. The enabling technologies for design, manufacturing, and operation of PSS are similar to those associated with Smart Manufacturing: IoT, PLM, and machine learning/data analytics.

10.5.3 Impact of PSS and Servitization on Manufacturing and Design

The new PSS business model of manufacturers with promise to continuously earning money as long as the PSS delivers value to the customer/user directly impacts the design and manufacturing of the products.

Again, we use two cases to understand these implications for design and manufacturing. Imagine the two business cases of a washing machine. On the one hand side we have a manufacturer selling their product to the customer and the customer owns it and maintains it and operates it himself or herself. After the product warranty expired, the manufacturer can only earn money with this customer again when the product breaks and has to be replaced with a new one. This principle has its own name – *planned obsolescence*. This is not very sustainable and counterproductive regarding customer satisfaction and loyalty. On the other hand, we have a PSS where the customer is guaranteed having the ability to wash. Maintenance, energy, and consumables (water, detergent, softener, etc.) are all taken care of with a subscription model. In this case, the manufacturer is earning the most money when the machine is not breaking down (lower maintenance cost) and is energy and resource efficient. This incentivizes the designer to use, e.g., higher quality and more expensive components (e.g., bearings and powertrains) that last longer and lead to more efficient operations – the opposite of planned obsolescence where components are designed to break after a certain time.

In the end, a PSS business model leads to better, more sustainable designs and manufacturing with less strain to the environment for two reasons: (i) access to usage data and information that inform better design and (ii) the implication of better design and use of better components/materials due to the chance to make more profit when the system is operational (vs. earning on maintenance/replacements before).

10.6 Cybersecurity in Manufacturing

We learned that the Fourth Industrial Revolution and Smart Manufacturing are based on connecting all manufacturing assets, collecting and analyzing manufacturing data on the shop floor and beyond. With this rapid digital transformation, new challenges that were previously not a priority for manufacturers emerge. An increasingly pressing challenge is cybersecurity. With increased connectivity between machines, shop floor systems such as Enterprise Resource Planning or Manufacturing Execution Systems, business units, and digital data repositories, have several critical threats for the safekeeping of company secrets, and safety of equipment and operators.

Cybersecurity is not new and has been a challenge in several domains for years. Cybersecurity is defined as "*the organization and collection of resources, processes, and structures used to protect cyberspace and cyberspace-enabled systems from occurrences that misalign de jure from de facto property rights*" (Craigen et al., 2014).

Manufacturing, however, possesses some distinct requirements that make cybersecurity in manufacturing stand out. In the last two years, several cybersecurity reports and frameworks were developed specifically designed for manufacturing needs. Two of them, the MForesight "Cybersecurity for Manufacturers" (Davis, 2017) and the NIST "Cybersecurity Framework Manufacturing Profile" (Stouffer et al., 2017) are good starting points if you are interested to dive deeper in the material.

In the next section, we will take a closer look at what makes cybersecurity threats possible in a manufacturing environment and discuss some examples on how this can impact the safety and well-being of the manufacturing company and its employees (Figure 10.11).

FIGURE 10.11   Cybersecurity increasing focus of manufacturers.

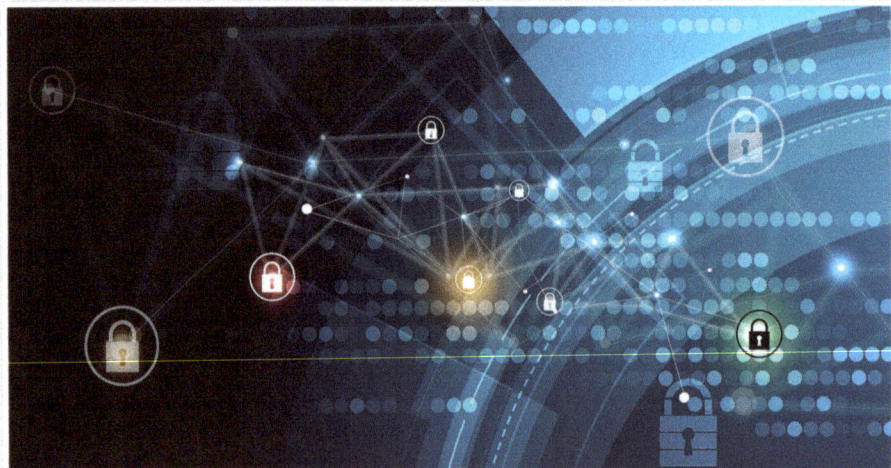

10.6.1 Cybersecurity Threats in Manufacturing

When we think about how much digital data exchange is happening every second in a connected factory following the Smart Manufacturing paradigm, it becomes apparent that there are several critical areas and possible cybersecurity threats that need to be addressed and monitored.

In order to structure the discussion, we created two clusters of possible cybersecurity threats, each with two levels of impact (see Table 10.1). It has to be understood that each of these categories and subcategories again includes a large number of possible instances of how cybersecurity can impact the company. We focus on the technical aspect of direct cybersecurity breaches on company systems and neglect the possible damage of an indirect cybersecurity breach, such as blackmailing an employee whose private account with compromising photos has been hacked. Nevertheless, social hacking is a real threat.

The examples depicted in Table 10.1 highlight that in manufacturing a cybersecurity breach can lead to not only financial losses of a company but, e.g., through manipulation of product data and process data, to bodily harm and significant damage. A more in-depth look into how cybersecurity threats can cause harm to manufacturing and design is provided by Wu et al. (2018).

Two examples illustrate the potential harm that can be done in more detail:

- Changing the process parameters of a manufacturing machine tool to cause critical machine failure – for example, increase the feed rate and depth of cut of a milling operation by a factor of 5 each. Or, as the famous computer worm "Stuxnet" that damaged centrifuges by infesting the Programmable Logic Controller of the machines in 2010.

- Another level 2 two-directional threat can be changing the design of a part essential for the structural integrity of an aircraft. This can be done in the CAD model itself, or after the CAM transformation in G-Code for machined parts or the STL file for 3D-printed structures. While the part will hold during minor loads, it might be maliciously designed to fail under large loads, e.g., during landing, where it will cause the most damage.

TABLE 10.1 Cybersecurity threats in a manufacturing environment

Cybersecurity threat	Sub-category	Examples
One-directional access	Level 1 - stealing historical data/information in form of documents, tables, etc.	• Company secrets can be revealed • Products may be copied and trade secrets used by competitors
	Level 2 - stealing real-time process data/information, e.g., CNC programs, sensor data, etc.	• Detailed manufacturing plans reveal information that can lead to insider trading • Products may be copied and manufacturing expertise as well
Two-directional access	Level 1 - stealing/manipulating/deleting historical data/information	• Required documents for certification can be manipulated cause problems with lawmakers and customers • Manipulated data can cause wrong prediction models when used to learn a machine learning model
	Level 2 - purposely manipulating CAD/CAM data, process parameters, etc.	• Can lead to quality issues, damaged products and machines • In the worst case, the manufactured products are not safe for use and can cause harm to users

© Harik/Wuest

10.6.2 Tools/Methods to Prevent Cybersecurity Threats

There are several works available on how to secure networks and communications. We will not focus on the general network security and obvious things like the use of a firewall, virus scanner, and secure passwords. What we will discuss is the manufacturing specific things that have to be kept in mind in order to avoid major leaks. However, this is by no means a comprehensive list, and in a rapidly changing domain like this, we suggest to always keep up to date with current threats and countermeasures.

First and foremost, manufacturers need to carefully assess what data, information, and knowledge are essential to their operation and put measures in place to safeguard these items carefully. This can mean a separate intranet that cannot be accessed from outside (and is not connected to the internet). This is a very tough decision to make, as having a separate infra-structure also limits the impact of this most likely important data, information, and knowledge. Therefore, this decision must be made by an interdisciplinary board (including key departments such as design, manufacturing, sales, maintenance, etc.) and signed off by the C-level executive team. Communicating the cybersecurity measures and training the workers is another key element of a successful cybersecurity strategy for manufacturers.

Many manufacturers are now digitally connected to their supply chain. This includes not only direct suppliers and customers but also suppliers' supplier, customers' customers, and third-party service providers. In a large company, like an automotive OEM, these supply chains can include hundreds of different organizations. It is essential to be aware who has access to what data and how far the integration is. Insisting on certification of all organizations involved in the data sharing is an important step towards cybersecurity. A new tool that promises major improvements is Blockchain. Blockchain is a way to creating a historical ledger that tracks transactions and exchange, and thus can establish trust and transparency of digital information exchange.

Within the organization, now that machine tools and other IoT devices are constantly connected, it is important to understand that the "smart fridge" in the kitchen area that is connected to the network might provide an entry point to the network. There is no 100% security guaranteed with the many new IoT devices that emerge in rapid succession today. Therefore, until cybersecurity of a device is transparent and guaranteed (ideally by a standard), it should not be connected to the network that includes sensitive data and/or manufacturing equipment.

10.7 Prognostics and Health Management (PHM)/Predictive Maintenance

When thinking of Smart Manufacturing and explaining the potential for industrial users, PHM and Predictive Maintenance are often used as a key application. One reason for this is that the benefits are very clear and easily understood by non-experts and manufacturing experts alike. In this section we will first discuss traditional maintenance approaches before we dive deeper into what PHM and predictive maintenance mean, and how they are different from the traditional methods.

Today, we distinguish five categories of maintenance strategies (or lack thereof): reactive maintenance, preventive maintenance, condition monitoring, predictive maintenance, and prescriptive maintenance. Reactive and preventive maintenance as well as partly condition monitoring are established, mainly analog maintenance approaches that are still widely employed in industry today. IoT-based condition monitoring, predictive maintenance, and prescriptive maintenance are core applications of Smart Manufacturing and Industry 4.0, and receive much attention in industry today as they offer unique benefits for the companies and service providers.

Reactive Maintenance is often understood as the most basic maintenance strategy or lack thereof. Basically, reactive maintenance implies that a company runs its equipment until it breaks before performing (emergency) maintenance and repair. The company is thus only reacting to the machines or equipment's failure, meaning that the timing of the equipment's downtime and availability of maintenance resources are not planned and reflected in the company's production planning. Therefore, this can have significantly impact the company's ability to fulfill orders on time and in the quality promised to customers. A way to avoid downtime is to have additional redundant production resources available which are a financial burden. This type of maintenance strategy is used mainly by smaller operations that do not have a dedicated maintenance team. Larger operations like an automotive OEM or its suppliers cannot operate on such a strategy as each hour of downtime, cause by equipment failure anywhere in the supply chain costs hundreds of thousands of dollars per hour.

A more advanced approach to maintenance is taken by a *Preventive Maintenance* strategy. In this case, maintenance is performed on a machine or equipment in regular intervals before any signs of problems. The idea is to make sure the machine is operational and stays operational to avoid unexpected breakdowns and downtime. While if performed correctly this strategy can prevent downtime successfully, the timing of maintenance intervals is critical. If the interval is chosen (too) short, the risk of breakdowns is reduced; however, the machine tool and maintenance resources are occupied during maintenance operations and cannot "produce." Furthermore, in case parts are replaced preventatively, these parts might still have been good for another XX hours of operation thus replacing them early is wasting valuable production resources. On the other hand, if intervals are chosen (too) long, the risk of breakdown and downtime increases. Choosing the right interval taking technical and economic factors into account is not trivial and also depends on several factors including

i. The products being produced (cheap, fast processing time mass-products - breakdowns not as critical; expensive, long-production time products, breakdowns critical if it can damage the part being produced)

ii. The customer (automotive OEM with Just-in-Sequence production, breakdowns not acceptable; retailer for, e.g., garden hoses made to stock, breakdowns more acceptable)

Condition monitoring is actually a core element of both predictive and preventive maintenance. However, in its basic form, it can also inform preventive maintenance. Condition monitoring describes the monitoring of a machine or piece of equipment regarding indicators for failure or problems. Often these indicators are certain process parameters like vibration, temperature (curves), and noise. By monitoring selected parameters and their change over time, the condition of the machine can be assessed. The data used for condition monitoring as part of predictive and prescriptive maintenance has to be remotely accessible and stored in a way that it is accessible and analyzable by machine learning and data analytics algorithms. The infrastructure necessary to accomplish that is IIoT as presented in the previous chapters.

Predictive Maintenance builds on the data gathered from condition monitoring. In predictive maintenance, the machine process parameter or combination of those parameters is monitored and labeled with respect to failure modes and time series. Through the use of machine learning and data analytics on the (historical) labeled data, prediction models are developed that are capable of picking up changes in the data set in real time and provide an advanced warning when the machine or equipment requires maintenance. Depending on the equipment, this warning can be far in advanced or very close to the actual breakdown. Generally, the closer the predicted failure the lower the uncertainty of the model. In any case, the integration of the prediction into the workflow of the maintenance department and production planners and other stakeholders is essential. In many cases, visualization of the data is key to ensure the right actions are taken after a future failure is identified (see Figure 10.12).

Examples of predictive maintenance applications are tool wear in cutting tools. In this case, the model is capable of predicting when a cutting tool needs to be replaced in order to either avoid it to break or to keep the required tolerances of the manufacturing process for the respective part. Another example is the maintenance of off-shore wind energy turbines.

FIGURE 10.12 Visualization of the predicted failure of component.

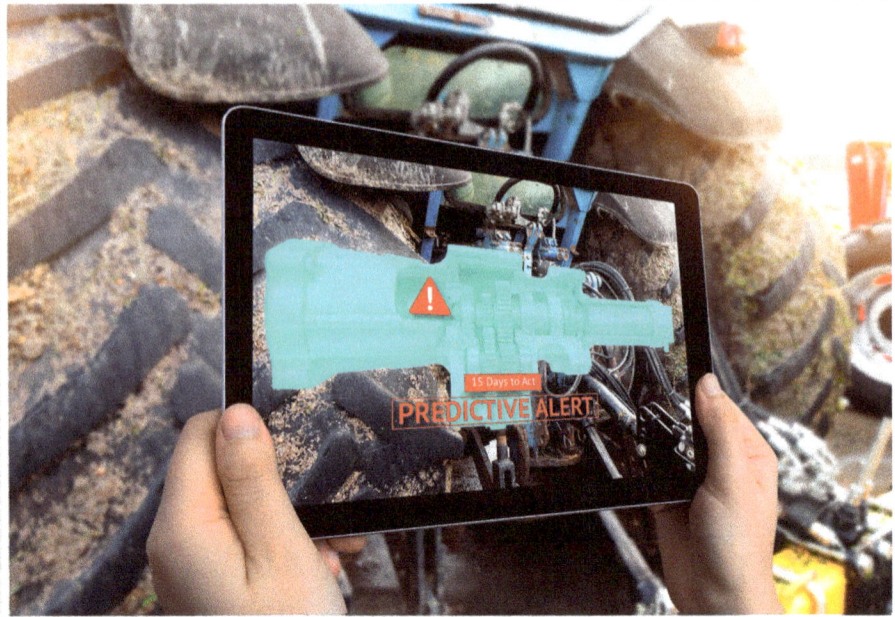

© Zapp2Photo/Shutterstock

In order to maintain these, a lot of resources are required and several constraints have to be dealt with. This includes the required experts, weather conditions, and special ships with cranes available. All in all, advanced warning and planning is essential to avoid major damage and downtime. The large bearings of these turbines are equipped with a large number of sensors, and the data is constantly communicated and analyzed to identify issues and predict time to failure in order to plan the maintenance operations accordingly. If successful, predictive maintenance juggles the need to prevent failure and downtime with the ability to utilize the maximum of the useful life of a machine or machine part.

Prescriptive Maintenance is sometimes seen as an instance of predictive maintenance. However, we believe the distinct difference of using analytics models not only to identify ("what") and predict potential failures ("when") but to identify causes ("how") and mitigate risks actively justifying elevating it to its own category. Prescriptive maintenance does not only monitor the condition of the equipment and predict potential failures using machine learning, it also includes different operational parameters and adaptations of the process into considerations to identify the ideal conditions to operate the equipment, as well as actively plans the maintenance strategy based on the data model and updates the predicted outcome in real time (Table 10.2).

TABLE 10.2 Comparison of different maintenance strategies

	Reactive maintenance	Preventive maintenance	Condition monitoring	Predictive maintenance	Prescriptive maintenance
Ease of implementation	++	+	o	−	−−
Avoid unplanned downtime of equipment	−−	o	+	++	++
Use of maintenance resources	+	−	+	++	++
Use of machine learning	−−	−	o	+	++
Based on data/IIoT	−−	−−	+	++	++
Real-time adaption of process	−−	−−	−	+	++

© Harik/Wuest

References

BMBF, "Industry 4.0," 2018, https://www.bmbf.de/de/zukunftsprojekt-industrie-4-0-848.html.

Boschert, S. and Rosen, R., "Digital Twin – The Simulation Aspect," in Hehenberger, P. and Bradley, D. (Eds.), *Mechatronic Futures: Challenges and Solutions for Mechatronic Systems and Their Designers* (Springer, 2016), 59–74.

Chand, S. and Davis, J.F., "What Is Smart Manufacturing," *Time Magazine Wrapper*, 2010, 28–33.

Craigen, D., Diakun-Thibault, N., and Purse, R., "Defining Cybersecurity," *Technology Innovation Management Review* 4, no. 10 (October 2014): 13-21.

Davis, J., "Cybersecurity for Manufacturers: Securing the Digitized and Connected Factory," Mforesight Report, 2017.

Davis, J., Edgar, T., Graybill, R., Korambath, P. et al., "Smart Manufacturing," *Annual Review of Chemical and Biomolecular Engineering* 6, no. 1 (2015): 141–160, doi:10.1146/annurev-chembioeng-061114-123255.

Harding, J.A., Shahbaz, M., Srinivas, and Kusiak, A., "Data Mining in Manufacturing: A Review," *Journal of Manufacturing Science and Engineering* 128, no. 4 (2006): 969, doi:10.1115/1.2194554.

Lang, S., "Durchgängige Mitarbeiterinformation zur Steigerung von Effizienz und Prozesssicherheit in der Produktion," Dissertation, Universität Erlangen-Nürnberg (Bamberg: Meisenbach Verlag, 2007).

Lee, J., Lapira, E., Bagheri, B., and Kao, H., "Recent Advances and Trends in Predictive Manufacturing Systems in Big Data Environment," *Manufacturing Letters* 1, no. 1 (2013): 38–41.

Loyer, J.-L., Henriques, E., Fontul, M., and Wiseall, S., (2016). "Comparison of Machine Learning Methods Applied to the Estimation of Manufacturing Cost of Jet Engine Components," *International Journal of Production Economics*, doi:10.1016/j.ijpe.2016.05.006.

McKinsey, "Smartening up with Artificial Intelligence (AI) - What's in It for Germany and Its Industrial Sector?," 2017, https://www.mckinsey.de/files/170419_mckinsey_ki_final_m.pdf.

Mittal, S., Kahn, M., Romero, D., and Wuest, T., "Smart Manufacturing: Characteristics, Technologies and Enabling Factors," *Part B: Journal of Engineering Manufacture* (2017), Online first, 1-20, doi:10.1177/0954405417736547 (open access).

Mizuno, H., Mori, Y., Taniguchi, Y., and Tsuji, H., "Data Queries Using Data Visualization Techniques," in *Systems, Man, and Cybernetics, 1997. IEEE International Conference on Computational Cybernetics and Simulation*, 1997, vol. 3, 2392-2396.

Monostori, L., "AI and Machine Learning Techniques for Managing Complexity, Changes and Uncertainties in Manufacturing," *Engineering Applications of Artificial Intelligence* 16, no. 4 (2003): 277–291, doi:10.1016/S0952-1976(03)00078-2.

Nilsson, N.J., *Introduction to Machine Learning* (Stanford, 2005).

Priore, P., de la Fuente, D., Puente, J., and Parreño, J., "A Comparison of Machine-Learning Algorithms for Dynamic Scheduling of Flexible Manufacturing Systems," *Engineering Applications of Artificial Intelligence* 19, no. 3 (2006): 247–255. doi:10.1016/j.engappai.2005.09.009

Samuel, A., "Some Studies in Machine Learning Using the Game of Checkers," *IBM Journal* 3, no. 3 (1959): 210–229.

Shirase, K. and Nakamoto, K., "Simulation Technologies for the Development of an Autonomous and Intelligent Machine Tool," *International Journal of Automation Technology* 7, no. 1 (2013): 6-15.

Smola, A. and Vishwanathan, S.V.N., *Introduction to Machine Learning* (Cambridge, UK: Cambridge University Press, 2008).

Stouffer, K., Zimmerman, T., Tang, C., Lubell, J. et al., "Cybersecurity Framework Manufacturing Profile," NIST Interagency/Internal Report (NISTIR)-8183, 2017.

Thoben, K.-D., Wiesner, S., and Wuest, T., ""Industrie 4.0" and Smart Manufacturing – A Review of Research Issues and Application Examples," *International Journal of Automation Technology* 11, no. 1 (2017): 4-19, doi:10.20965/ijat.2017.p0004.

Vandermerwe, S. and Rada, J., "Servitization of Business: Adding Value by Adding Services," *European Management Journal* 6, no. 4 (1988): 314–324.

Wallace, E. and Riddick, F., *Panel on Enabling Smart Manufacturing* (USA: State College, 2013).

Wu, D., Ren, A., Zhang, W., Fan, F. et al., "Cybersecurity for Digital Manufacturing," *Journal of Manufacturing Systems* (2018), doi:10.1016/j.jmsy.2018.03.006.

Wuest, T., *Identifying Product and Process State Drivers in Manufacturing Systems Using Supervised Machine Learning*, Springer theses (New York/Heidelberg: Springer-Verlag, 2015).

Wuest, T., Hribernik, K., and Thoben, K.-D., "Accessing Servitisation Potential of PLM Data by Applying the Product Avatar Concept," *Production Planning & Control* 26, no. 14-15 (2015): 1198-1218, doi:10.1080/09537287.2015.1033494.

Wuest, T., Weimer, D., Irgens, C., and Thoben, K.-D., Machine Learning in Manufacturing: Advantages, Challenges and Applications. *Production & Manufacturing Research*, 4, no. 1 (2016): 23-45, doi:10.1080/21693277.2016.1192517.

Dr. Ramy Harik, a Fulbright Scholar, is an Associate Professor in the Department of Mechanical Engineering at the **University of South Carolina** (USC) and a resident researcher at the **McNAIR Aerospace Center**. He is an affiliated research scientist in Automated Fiber Placement at the National Aeronautics and Space Administration **(NASA) Langley** in the ISAAC team. His education joins Mechanical Engineering (Master's of Engineering) from the **Lebanese University ULFGII** (Roumieh, Lebanon) and Automated Manufacturing Engineering Technology (Master's of Science) and Industrial/Mechanical Engineering (Doctor of Philosophy) from the **University of Lorraine** (Nancy, France).

Dr. Harik has approximately taught Manufacturing Processes **over 20 times on three different continents!** That does not include other courses such as Computer Aided Manufacturing, Prototyping, Advanced Manufacturing, Quality Control, Ergonomics, and other manufacturing related topics. This multicultural immersion in different education environments and observation of multiple educational setups enabled him to perform continuous improvement on his teaching delivery methods. **This book represents the culmination of these teaching experiences** and interactions with the students of the University of Lorraine, Lebanese American University, and USC, his current tenure home. Since 2016, his average student rating (filled by over 300 students over 5 semesters) taking the manufacturing class is 4.96, 4.99, 4.97, 5, and 5. This demonstrates his passion to the students and the topic, leading him to receive the **2016 Outstanding Young Educator Award** from Pi Tau Sigma Honor Society at USC and the **2018 Samuel Litman Distinguished Professor Award** from the College of Engineering and Computing at USC.

Dr. Harik has created endless undergraduate and graduate programs to complement the in-class learning experience for students, such as the **McNAIR Junior Fellows** Undergraduate Research program at USC, now in its fourth year with over 50 fellows. These numerous activities (from over 100 field trips to industry to ABET Accreditation Platforms) demonstrate his fundamental belief in the education of the whole person, in and out of the classroom.

Dr. Harik has over **$3 million** in funding from NASA, Boeing, SCDoC, Toray, Fokker, SCRA, and other agencies. He held visiting positions at Purdue University and NASA Langley (USA) and Ecole Centrale de Nantes and UTC Compiegne (France). He has published over 100 peer-reviewed articles in international journals and conferences with over 900 citations to date. Dr. Harik was the chair of CAD'10 and the PLM'16 International Conferences.

Dr. Thorsten Wuest serves as Assistant Professor at West Virginia University and J. Wayne and Kathy Richards Faculty Fellow in Engineering at the WVU's Benjamin M. Statler College of Engineering and Mineral Resources. Dr. Wuest earned a Ph.D. in Production Engineering (Summa Cum Laude), a Master's in Industrial Engineering and Management, both University of Bremen, in Germany, and graduated top of his class in the International Business Master's program at the AUT University, New Zealand. During his Ph.D. he spent ten months as a visiting scholar at the University of Southern California, USA funded by the DAAD.

Dr. Wuest's research focusses on Smart and Advanced Manufacturing, Industry 4.0, Machine Learning, Hybrid Analytics, as well as Product Service Systems. Dr. Wuest gave invited talks in seven countries and published over 100 peer-reviewed articles in international archival journals and conferences and serves as a reviewer for many. In addition to publishing his work in the premier academic outlets of his field, Dr. Wuest was featured by Forbes, Futurism, World Economic Forum, CBC Radio, and World Manufacturing Forum, among others. He is a member of the Editorial Board for the *Journal of Manufacturing Systems*, World Manufacturing Forum, and an Associate Editor for the *Int'l Journal of Manufacturing Research*. He is involved in several professional societies, including IFIP Working Group 5.1 and 5.7, IFAC TC 5.3, IISE, SME, and CIRP.

Dr. Wuest is very passionate about advising and teaching graduate and undergraduate students. He recently received the WVU Statler College of Engineering and Mineral Resources "Outstanding Teacher of the Year 2019" Award. Currently, he teaches Manufacturing Processes for industrial, mechanical, and aerospace engineers as well as New Product and Service Development for industrial engineers. He is an inaugural IDEA Fellow in the office of the Provost of West Virginia University to further entrepreneurship education and mindset across the campus and develop a new, interdisciplinary course on Product Service Systems receiving outstanding student evaluations.

For more resources including online quizzed, homework, YouTube videos and more, please visit: www.Introtomanufacturing.com

For faculty resources including PowerPoints, exams, extra exercises and more, please email us at: Ramy.harik@gmail.com

9780768093278